Geometry

**LARSON
BOSWELL
STIFF**

Applying • Reasoning • Measuring

Basic Skills Workbook: Diagnosis and Remediation Teacher's Edition

The Basic Skills Workbook provides material you can use to review and practice basic prerequisite skills for Geometry. A diagnostic test is provided to help you determine which topics in the workbook need to be reviewed. The Teacher's Edition includes the student workbook plus assessment materials and answers.

McDougal Littell
A HOUGHTON MIFFLIN COMPANY
Evanston, Illinois • Boston • Dallas

Contributing Author

The authors wish to thank Glenda Haynie for her contribution to the Basic Skills Workbook: Diagnosis and Remediation.

A Note to Teachers

The Student Workbook includes about a month of instruction on topics that are prerequisites for Geometry. Each of the five topics comprises four brief lessons with exercises, and each lesson is accompanied by a separate sheet of Quick Check exercises that review the previous lesson and a sheet of extra practice to follow the lesson.

To help you determine which topics will benefit your students, a diagnostic test is provided at the beginning of the Student Workbook. Individual items on the diagnostic test are keyed to each lesson.

The Teacher's Edition also includes a brief assessment following each topic, a cumulative assessment following the last topic, and a complete set of answers.

ISBN: 0-618-02076-4

456789-VEI- 04 03 02

Contents

Diagnostic Test

For use before Topic 1

Adding Rational Numbers (Topic 1, Lesson 1, pages 1–5)

Find the absolute value.

1. $|6.73|$ 2. $\left|-\frac{4}{9}\right|$

3. $|-0.33|$ 4. $\left|2\frac{5}{12}\right|$

5. $|0|$ 6. $\left|-\frac{21}{16}\right|$

Find the sum.

7. $3.55 + 1.24$ 8. $(-1.18) + 2.31$

9. $\left(-\frac{1}{6}\right) + \frac{1}{6}$ 10. $\left(-\frac{1}{6}\right) + \left(-\frac{1}{6}\right)$

11. $\frac{13}{4} + \left(-\frac{13}{4}\right)$ 12. $12.3 + (-28.004)$

13. $(-16.81) + (-31.5)$ 14. $(-2.\overline{67}) + 2.\overline{67}$

15. $19\frac{5}{8} + 0$ 16. $\left(-\frac{5}{11}\right) + \left(\frac{19}{11}\right) + \left(-\frac{12}{11}\right)$

17. $(-12) + 8.12 + 12$ 18. $\left(-9\frac{2}{5}\right) + \left(-\frac{1}{4}\right) + 20$

Subtracting Rational Numbers (Topic 1, Lesson 2, pages 6–9)

Find the opposite of the number.

19. 16.22 20. $-\frac{5}{6}$

21. -5.0

Find the difference.

22. $15.63 - 0.55$

23. $(-9.02) - 6.37$

24. $16.23 - (-7.02)$

25. $\left(-\frac{11}{12}\right) - \left(-\frac{5}{12}\right)$

26. $\frac{3}{16} - \left(-\frac{7}{16}\right)$

27. $\left(-\frac{7}{9}\right) - \left(\frac{1}{9}\right)$

28. $(-3.\overline{2}) - (-3.\overline{2})$

29. $(-22) - (-18.19)$

1. _____
2. _____
3. _____
4. _____
5. _____
6. _____
7. _____
8. _____
9. _____
10. _____
11. _____
12. _____
13. _____
14. _____
15. _____
16. _____
17. _____
18. _____
19. _____
20. _____
21. _____
22. _____
23. _____
24. _____
25. _____
26. _____
27. _____
28. _____
29. _____

Diagnostic Test

For use before Topic 1

30. $\left(-\frac{2}{5}\right) - \left(-\frac{16}{25}\right)$ **31.** $\left(-\frac{5}{18}\right) - \left(\frac{7}{15}\right)$

32. $42\frac{2}{3} - 0$ **33.** $1 - \left(\frac{27}{13}\right)$

Multiplying and Dividing Rational Numbers

(Topic 1, Lesson 3, pages 10–15)

Find the reciprocal of the number.

34. -46 **35.** $\frac{9}{10}$ **36.** $2\frac{4}{9}$

Find the product or quotient.

37. $\left(-\frac{4}{5}\right)\left(-\frac{1}{5}\right)$ **38.** $\left(-\frac{5}{6}\right) \div \left(-\frac{1}{2}\right)$

39. $\left(-\frac{1}{2}\right)(-2)$ **40.** $(50.16)(-1.5)$

41. $(-16.24) \div (-0.4)$ **42.** $\left(-1\frac{3}{5}\right) \div \left(-2\frac{1}{10}\right)$

43. $\left(-\frac{4}{9}\right)\left(5\frac{1}{2}\right)$ **44.** $(-36) \div \frac{1}{6}$

45. $2626 \div 26.26$

Evaluate the expression.

46. $\left(-\frac{5}{16}\right)(-1) \div \left(\frac{5}{16}\right)$

47. $\left(\frac{3}{8}\right)(0)\left(-\frac{8}{3}\right)$

48. $10 \div \frac{7}{10} \cdot \left(-\frac{7}{10}\right)$

Order of Operations (Topic 1, Lesson 4, pages 16–21)

Evaluate the expression.

49. $-3.2 + 6(-1.3)$

50. $5 \cdot \left(\frac{1}{4}\right) \div 6 \cdot (-2)$

51. $(0.2)^3 - (1.2)(1.4)$

52. $\left(-\frac{1}{2}\right)^2\left(-\frac{3}{8}\right) + 4$

53. $\left(-\frac{1}{2} - \frac{1}{3}\right)^2 - 1$

54. $\left(\frac{20}{9}\right) \div \left(-\frac{20}{9}\right) \cdot 5$

55. $2.1 + (0.6)^2 - 4$

56. $12 - 6\left(\frac{1}{3}\right)^2$

57. $\left(1\frac{1}{8} + 2\right) \div \left(3 - 1\frac{7}{8}\right)$

58. $26 - (18.1 - 0.8)$

30. _____

31. _____

32. _____

33. _____

34. _____

35. _____

36. _____

37. _____

38. _____

39. _____

40. _____

41. _____

42. _____

43. _____

44. _____

45. _____

46. _____

47. _____

48. _____

49. _____

50. _____

51. _____

52. _____

53. _____

54. _____

55. _____

56. _____

57. _____

58. _____

59. $19\frac{1}{2} - \left(1 - \frac{1}{4}\right)$

60. $(-18) \div (0.3) \cdot (0.4) - 16$

Evaluating Expressions (Topic 2, Lesson 1, pages 23–28)

Evaluate the expression when $x = -8$ and $y = \frac{3}{4}$.

61. xy

62. $-x + 8y$

63. $\frac{1}{8}x - \frac{4}{3}y$

64. $\frac{y}{3}$

65. $-|x|$

66. $(x + 3)^2$

67. $|x + 1| - 16y$

68. $-x^2 + y^2$

69. $|x| + |y|$

Simplifying Linear Expressions in One Variable
(Topic 2, Lesson 2, pages 29–32)

Simplify by combining like terms.

70. $7x + 3x - 15$

71. $-b - 6b$

72. $(3y + 4)(-1) + y$

73. $2(c - 3) - 3(c + 1)$

74. $5(6 - 2d) - d$

75. $-\frac{11}{4}y - \frac{1}{6}y$

76. $\frac{4}{9}m + m + \frac{1}{3}m$

77. $5\frac{1}{6} + 2t - 1$

78. $19x + (10 - x)3$

Simplifying Expressions in Two Variables
(Topic 2, Lesson 3, pages 33–36)

Simplify by combining like terms.

79. $15x + y + x$

80. $xy - xy + 3y$

81. $20a^2 + 2a - 7a^2$

82. $x^2y + 5xy - 3xy - 1$

83. $18x^3 + 16x^2 + 3x$

84. $119x - y + \frac{5}{6}y$

59. _____

60. _____

61. _____

62. _____

63. _____

64. _____

65. _____

66. _____

67. _____

68. _____

69. _____

70. _____

71. _____

72. _____

73. _____

74. _____

75. _____

76. _____

77. _____

78. _____

79. _____

80. _____

81. _____

82. _____

83. _____

84. _____

Diagnostic Test

For use before Topic 1

85. $-\frac{1}{2}s^2t + \frac{1}{4}st^2 - \frac{1}{6}st^2$ **86.** $(5h - k) - (-k + h)$

87. $-b^2 + 1.6b + (-b)$

Using Properties with Expressions (Topic 2, Lesson 4, pages 37–40)

Identify the mathematical property illustrated in the identity.

88. $5 - x^2 = -x^2 + 5$

89. $-6x - (12 - 3x) = -6x - 12 + 3x$

90. $-\frac{1}{2}ab + \frac{1}{2}ab = 0$

91. $(12t - 6y)0 = 0$

Show the steps and state the mathematical properties used in simplifying the expression.

92. $12(x - 3) - 6(5x - 1)$ **93.** $\left(\frac{1}{5} + y + \frac{2}{5} - y\right)(2x + y + 8x)$

94. $14e + f - 7e + f - 2f$ **95.** $29 - 5(x - 3)$

Solving Equations (Topic 3, Lesson 1, pages 42–47)

Solve the equation.

96. $\frac{p}{7} = -8$ **97.** $x + 13 = -9$ **98.** $y - 10 = -19$

99. $16s = -48$ **100.** $\frac{p}{21} = 0$ **101.** $14x = x$

102. $8.25m - m = 1.45$ **103.** $-t + 63 = 21$

104. $3 + 5(4z + 6) = 2z - 3$

Solving Inequalities (Topic 3, Lesson 2, pages 48–53)

Check to see if the number given is a solution of the inequality.

105. $-x > 24, -25$

106. $5y \le 4y, \frac{3}{2}$

107. $11 - (c + 2) < 9, 0$

Solve the inequality.

108. $\frac{x}{4} \le -3$ **109.** $5m > -1$

110. $n - 4 < -11$ **111.** $14x < x$

112. $-7x \le 35$ **113.** $-x + 2 \ge -2$

85. _____
86. _____
87. _____
88. _____
89. _____
90. _____
91. _____
92. _____
93. _____
94. _____
95. _____
96. _____
97. _____
98. _____
99. _____
100. _____
101. _____
102. _____
103. _____
104. _____
105. _____
106. _____
107. _____
108. _____
109. _____
110. _____
111. _____
112. _____
113. _____

NAME _____   DATE _____

Diagnostic Test

For use before Topic 1

114. $15 + 3(x - 2) > -9$ **115.** $4(-3)^2 + p > 50$

116. $\frac{1}{12}r - 6 \geq \frac{1}{12}$

Proportions (Topic 3, Lesson 3, pages 54–58)

Solve the proportion.

117. $\dfrac{x}{8} = \dfrac{3}{4}$ **118.** $\dfrac{15}{x} = \dfrac{5}{6}$ **119.** $\dfrac{13}{65} = \dfrac{5}{k}$

120. $\dfrac{9}{a} = \dfrac{36}{-20}$ **121.** $\dfrac{17}{100} = \dfrac{8.5}{x}$ **122.** $\dfrac{100}{145} = \dfrac{c}{29}$

123. $\dfrac{x}{28} = \dfrac{x - 1}{50}$ **124.** $\dfrac{s + 3}{s} = \dfrac{5}{4}$ **125.** $\dfrac{9}{5} = \dfrac{w + 21}{w + 5}$

Solving Problems (Topic 3, Lesson 4, pages 59–63)

126. *Final Exam Grade*   Your final average in algebra class is 75% of your semester average plus 25% of your final exam. If your semester average is 90 points, and your final average is 92, what is your final exam grade?

127. *Book Club*   You join a book club. The beginning membership is $15. Books are all the same price, which includes tax. What is the price of a book if you buy 4 books when you send in your membership and write a check for $66.80?

Drawing and Measuring Angles (Topic 4, Lesson 1, pages 65–69)

Use a protractor to find the measure of $\angle A$ to the nearest degree. Then identify the angle as acute, right, obtuse, or straight.

128.

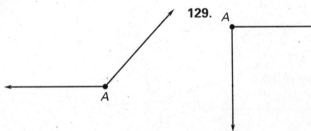

129.

130.

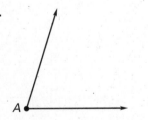

114.	_____
115.	_____
116.	_____
117.	_____
118.	_____
119.	_____
120.	_____
121.	_____
122.	_____
123.	_____
124.	_____
125.	_____
126.	_____
127.	_____
128.	_____
129.	_____
130.	_____

Diagnostic Test

For use before Topic 1

Use a protractor to draw an angle with the degree measurement given and label the angle appropriately.

131. $m\angle H = 85°$ **132.** $m\angle ABC = 121°$ **133.** $m\angle A = 164°$

Polygons (Topic 4, Lesson 2, pages 70–74)

Classify the triangle by its sides and angles.

134. 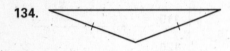   **135.**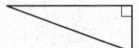

136.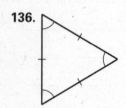

Give the name that best describes each quadrilateral.

137. 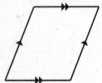   **138.**   **139.**

Circles (Topic 4, Lesson 3, pages 75–78)

Use a compass to draw a circle with the given radius or diameter. Then complete the statement.

140. radius = 7 cm   **141.** diameter = 1 in.   **142.** radius = 30 mm
    diameter = $\underset{?}{\_\_}$ radius = $\underset{?}{\_\_}$ diameter = $\underset{?}{\_\_}$

Solids (Topic 4, Lesson 4, pages 79–83)

In Exercises 143-146, use the prism at the right.

143. List the edges.

144. List the vertices.

145. List the faces.

146. List the bases.

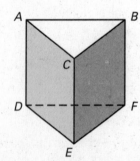

131.	_____
132.	_____
133.	_____
134.	_____
135.	_____
136.	_____
137.	_____
138.	_____
139.	_____
140.	_____
141.	_____
142.	_____
143.	_____
144.	_____
145.	_____
146.	_____

DIAGNOSTIC TEST CONTINUED

Diagnostic Test

For use before Topic 1

Identify the solid.

147. 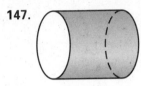   **148.** 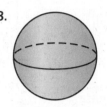   **149.**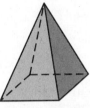

150.    **151.** 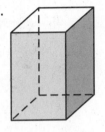   **152.**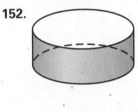

Plotting Points (Topic 5, Lesson 1, pages 86–90)

Plot and label each of the following points on the coordinate plane.

153. $A(0, 6)$ **154.** $B(-5, 0)$ **155.** $C(2, 7)$

156. $D(-1, -4)$ **157.** $E(3, -5)$ **158.** $F(-4, 2)$

Give the coordinates of each point on the graph and the quadrant or axis in which it lies.

159. A

160. B

161. C

162. D

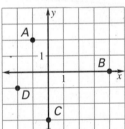

Graphing Equations (Topic 5, Lesson 2, pages 91–95)

Graph the linear equation on graph paper.

163. $y = x + 7$ **164.** $y = -2x + 6$ **165.** $3x - y = 0$

166. $x + y = 11$ **167.** $y = -4$ **168.** $4x + 5y = 20$

Figures in a Plane (Topic 5, Lesson 3, pages 96–100)

Graph each set of points, and then connect them in order with segments to form a polygon. Then name the polygon.

169. $(-6, -2), (2, -2), (2, 1), (-6, 1)$

170. $(5, -2), (-5, 0), (0, 1)$

147. _____
148. _____
149. _____
150. _____
151. _____
152. _____
153. _____
154. _____
155. _____
156. _____
157. _____
158. _____
159. _____
160. _____
161. _____
162. _____
163. _____
164. _____
165. _____
166. _____
167. _____
168. _____
169. _____
170. _____

Geometry
Basic Skills Workbook: Diagnosis and Remediation

Diagnostic Test

For use before Topic 1

171. Transform the polygon in Exercise 169 by adding 5 to each x-coordinate.

172. Transform the polygon in Exercise 170 by replacing each y-coordinate with $-y$.

Perimeter and Area (Topic 5, Lesson 4, pages 101–106)

The points are the vertices of a figure in the coordinate plane. Plot the points and find the perimeter and area of the figure.

173. $(-1, -1), (4, -1), (4, 2), (-1, 2)$

174. $(-5, 0), (-1, 0), (-1, -4), (-5, -4)$

175. $(1, -1), (3, -1), (3, 7), (1, 7)$

176. $(0, 1), (0, 3), (-4, 3), (-4, 1)$

171._____

172._____

173._____

174._____

175._____

176._____

177._____

178._____

179._____

Divide and/or rearrange the shaded figure to find its area.

177.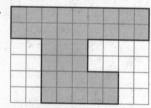

178.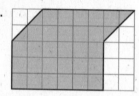

179.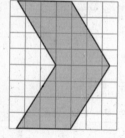

Geometry
Basic Skills Workbook : Diagnosis and Remediation

NAME _____ DATE _____

Adding Rational Numbers

GOALS **Add rational numbers.**

> Rational numbers can be added using the concept of absolute value and the rules and properties of addition.

Terms to Know

Example/Illustration

Terms to Know	Example/Illustration
Rational Number any number that can be written in the form $\frac{a}{b}$ where a and b are integers (Terminating and repeating decimals are rational since they can be written in fractional form.)	$\frac{1}{2}$ $3\frac{2}{3}$ 0 -5 0.69 $0.\overline{3}$ *Note*: $3\frac{2}{3} = \frac{11}{3}$ $0 = \frac{0}{1}$ $-5 = \frac{-5}{1}$ $\quad\quad 0.69 = \frac{69}{100}$ $0.\overline{3} = \frac{1}{3}$ π and $\sqrt{2}$ are not rational numbers since they are nonrepeating, nonterminating decimals and cannot be written in fractional form. They are irrational numbers.
Absolute value on a number line, the distance from the number to 0 (The symbol $|x|$ is read "the absolute value of x.")	$\left|\frac{1}{2}\right| = \left|-\frac{1}{2}\right| = \frac{1}{2}$ $\frac{1}{2}$ and $-\frac{1}{2}$ are both $\frac{1}{2}$ unit from 0.

Understanding the Main Ideas

The rules of addition are the guidelines for adding all signed numbers without a number line. You will use these rules when adding rational numbers.

> **Rules of Addition**
>
> **To add two numbers with the same sign:**
> **STEP 1:** Add their absolute values.
> **STEP 2:** Attach the common sign.
>
> $\quad$ **Example:** $-\frac{1}{2} + \left(-\frac{3}{2}\right)$
>
> $\quad$ Step 1 $\rightarrow$ $\left|-\frac{1}{2}\right| + \left|-\frac{3}{2}\right| = \frac{4}{2}$ or 2 $\quad$ **Step 2** $\rightarrow$ -2
>
> **To add two numbers with opposite signs:**
> **STEP 1:** Subtract the smaller absolute value from the larger absolute value.
> **STEP 2:** Attach the sign of the number with the larger absolute value.
> $\quad$ **Example:** $1.05 + (-3.12)$
> $\quad$ Step 1 $\rightarrow$ $|-3.12| - |1.05| = 2.07$ $\quad$ **Step 2** $\rightarrow$ -2.07

(continued)

NAME _____ DATE _____

Adding Rational Numbers

EXAMPLE 1

Use the rules of addition to find each sum.

a. $-0.54 + (-4.26)$

b. $\left(-\frac{5}{6}\right) + \left(\frac{1}{3}\right)$

SOLUTION

a. STEP 1: $|-0.54| + |-4.26| = 0.54 + 4.26$ Add absolute values.

$$= 4.80$$

STEP 2: -4.80 Attach the common sign.

b. First rewrite fractions with a common denominator to determine which has the larger absolute value.

$$\left(-\frac{5}{6}\right) + \left(\frac{1}{3}\right) = \left(-\frac{5}{6}\right) + \left(\frac{2}{6}\right)$$

STEP 1: $\left|-\frac{5}{6}\right| - \left|\frac{2}{6}\right| = \frac{5}{6} - \frac{2}{6}$ Subtract the smaller absolute value.

$$= \frac{1}{2}$$

STEP 2: $-\frac{1}{2}$ Attach negative sign.

Find the sum.

1. $\left(-\frac{1}{4}\right) + \left(-\frac{3}{4}\right)$ **2.** $6.92 + (-1.21)$ **3.** $\left(-7\frac{2}{5}\right) + \left(\frac{1}{5}\right)$

The properties of addition are also useful when adding rational numbers. They are listed here for you to review.

Properties of Addition

Commutative Property

The order in which two numbers are added does not change the sum.

$a + b = b + a$

Example: $\frac{1}{2} + \left(-\frac{1}{4}\right) = \left(-\frac{1}{4}\right) + \frac{1}{2}$

Associative Property

The way you group three numbers when adding does not change the sum.

$(a + b) + c = a + (b + c)$

Example: $\left[\frac{1}{2} + \left(-\frac{1}{4}\right)\right] + \frac{1}{3} = \frac{1}{2} + \left[\left(-\frac{1}{4}\right) + \frac{1}{3}\right]$

(continued)

Geometry
Basic Skills Workbook: Diagnosis and Remediation

Adding Rational Numbers

Identity Property
The sum of a number and 0 is the number.
$a + 0 = a$ **Example:** $7.5 + 0 = 7.5$

Property of Zero
The sum of a number and its opposite is 0.
$a + (-a) = 0$ **Example:** $2.637 + (-2.637) = 0$

EXAMPLE 2

Use the properties of addition to find each sum.

a. $(-0.5) + 3.1 + 0.5$

b. $\left(-6\frac{1}{3}\right) + 7 + \left(-3\frac{2}{3}\right)$

SOLUTION

a. $(-0.5) + 3.1 + 0.5 = 3.1 + (-0.5) + 0.5$ Use commutative property.

$= 3.1 + [(-0.5) + 0.5]$ Use associative property.

$= 3.1 + 0$ Use property of zero.

$= 3.1$ Use identity property.

b. $\left(-6\frac{1}{3}\right) + 7 + \left(-3\frac{2}{3}\right) = \left(-6\frac{1}{3}\right) + \left(-3\frac{2}{3}\right) + 7$ Use commutative property.

$= -10 + 7$ Add mixed numbers.

STEP 1: $|-10| - |7| = 10 - 7$ Subtract the smaller absolute value.

$= 3$

STEP 2: -3 Attach negative sign.

Find the sum.

4. $\dfrac{6}{7} + 1 + \left(-\dfrac{6}{7}\right)$ **5.** $-7.8 + 3 + (-2.2)$ **6.** $(-100 + 25.98) + 100$

Mixed Review

Find the sum of the integers.

7. $(-12) + 4$ **8.** $(-8) + (-12)$

9. $6 + 25 + (-6)$ **10.** $4 + (-3) + 6 + (-17)$

Geometry
Basic Skills Workbook: Diagnosis and Remediation

NAME _____ DATE _____

Quick Check

Review of real numbers and their properties

Standardized Testing Quick Check

1. If x is negative, $|x|$ equals

   A. x.

   B. $-x$.

   C. $\pm x$.

   D. $\sqrt{x}$.

   E. none of these

2. The statement $6 + 4 + (-6) = 4 + 6 + (-6)$ is an example of which property?

   A. Commutative Property

   B. Associative Property

   C. Identity Property

   D. Property of Zero

   E. none of these

Homework Review Quick Check

Find the sum.

3. $(-8) + (-3)$

4. $3 + (-8)$

5. $2 + (-35) + (-2)$

Geometry
Basic Skills Workbook: Diagnosis and Remediation

NAME _____ DATE _____

Practice

For use with Lesson 1.1: Adding Rational Numbers

Find the absolute value.

1. $|9.37|$

2. $\left|-\dfrac{1}{7}\right|$

3. $|0|$

4. $\left|3\dfrac{4}{9}\right|$

5. $|-0.6\overline{1}|$

6. $\left|-\dfrac{13}{5}\right|$

Find the sum.

7. $4.27 + 0.13$

8. $(-4.27) + 0.13$

9. $4.27 + (-0.13)$

10. $\left(-\dfrac{1}{2}\right) + \left(-\dfrac{1}{2}\right)$

11. $\dfrac{1}{2} + \left(-\dfrac{1}{2}\right)$

12. $\left(-\dfrac{3}{8}\right) + \dfrac{1}{8}$

13. $\dfrac{11}{3} + \left(-\dfrac{11}{3}\right)$

14. $\left(-\dfrac{11}{3}\right) + \left(-\dfrac{11}{3}\right)$

15. $(-1.\overline{34}) + 1.\overline{34}$

16. $29.1 + 37.001$

17. $(-135) + 1.043$

18. $(-54.98) + (-1289.1)$

19. $\left(-\dfrac{1}{5}\right) + \left(-\dfrac{3}{10}\right)$

20. $\left(-\dfrac{3}{8}\right) + \dfrac{3}{7}$

21. $\left(-\dfrac{3}{8}\right) + \dfrac{5}{11}$

22. $(-17) + \dfrac{1}{4}$

23. $11\dfrac{2}{9} + \dfrac{1}{9}$

24. $\left(-13\dfrac{5}{6}\right) + \left(-51\dfrac{1}{3}\right)$

25. $36\dfrac{41}{49} + 0$

26. $(-0.\overline{8}) + (-1)$

27. $1 + \left(-\dfrac{20}{3}\right)$

28. $(-8) + 4.63 + 8$

29. $\left(-\dfrac{45}{7}\right) + \dfrac{11}{7} + \left(-\dfrac{15}{7}\right)$

30. $51 + 1.\overline{5} + (-1.\overline{5})$

31. $95.4 + (-39.37) + 4.6$

32. $(-45.08) + (-54) + (-0.92)$

33. $\left(-4\dfrac{2}{3}\right) + \left(-\dfrac{1}{12}\right) + 10$

34. *Checkbook* Darrell is balancing his checkbook. He has deposits of $324.78 and $1004.61. He wrote three checks for $123.85, $62.00, and $46.79. What is his ending balance if his starting balance was zero?

35. *Molding* Jim needs to buy molding for a new door in his home. He needs $83\dfrac{3}{8}$ inches for each side and $41\dfrac{1}{2}$ inches for the top. How many inches of molding does he need to buy?

Subtracting Rational Numbers

GOAL **Subtract rational numbers.**

Adding the opposite of a number is equivalent to subtracting the number.

Terms to Know	Example/Illustration
Opposites two numbers whose sum is zero	-3.69 is the opposite of 3.69 since $-3.69 + 3.69 = 0$.

Understanding the Main Ideas

Subtracting rational numbers that involve negative numbers can be summarized with the following rule.

Subtraction Rule

To subtract b from a, add the opposite of b to a.

$$a - b = a + (-b)$$

Example: $\dfrac{3}{4} - \dfrac{1}{4} = \dfrac{3}{4} + \left(-\dfrac{1}{4}\right)$

The result is the difference of a and b.

EXAMPLE 1
_____

Rewrite each subtraction expression as an addition expression.

a. $(-43.89) - (-6.004)$ **b.** $\dfrac{7}{5} - \dfrac{9}{5}$

SOLUTION

a. $(-43.89) - (-6.004) = (-43.89) + (6.004)$ Add the opposite of -6.004.

b. $\dfrac{7}{5} - \dfrac{9}{5} = \dfrac{7}{5} + \left(-\dfrac{9}{5}\right)$ Add the opposite of $\dfrac{9}{5}$.

Rewrite the subtraction expression as an addition expression.

1. $4 - 5.38$ **2.** $\left(-\dfrac{1}{8}\right) - \left(-\dfrac{1}{4}\right)$ **3.** $\left(-10\dfrac{1}{11}\right) - \dfrac{6}{11}$

EXAMPLE 2
_____

Find each difference.

a. $-0.54 - (-4.26)$ **b.** $\left(-\dfrac{5}{6}\right) - \dfrac{1}{3}$

SOLUTION

a. $-0.54 - (-4.26) = -0.54 + (4.26)$ Add the opposite of -4.26.

$= 3.72$ Use rules of addition.

(continued)

Geometry
Basic Skills Workbook: Diagnosis and Remediation

NAME _____ DATE _____

Subtracting Rational Numbers

b. $\left(-\frac{5}{6}\right) - \frac{1}{3} = \left(-\frac{5}{6}\right) + \left(-\frac{1}{3}\right)$ Add the opposite of $\frac{1}{3}$.

$\qquad = \left(-\frac{5}{6}\right) + \left(-\frac{2}{6}\right)$ Rewrite with common denominator.

$\qquad = -\frac{7}{6}$ Use rules of addition.

Find the difference.

4. $\left(-\frac{1}{8}\right) - \left(-\frac{3}{4}\right)$ **5.** $0.912 - (-1.221)$ **6.** $-37\frac{1}{3} - 9\frac{2}{3}$

Expressions containing more than one subtraction can also be evaluated by adding the opposite. To do this, use the left-to-right rule for order of operations.

EXAMPLE 3

Evaluate each expression.

a. $6 - 8 - 9.3 - 3.1$ **b.** $-\frac{9}{8} + \frac{1}{4} - \left(-\frac{5}{4}\right) - \frac{1}{4}$

SOLUTION

a. $6 - 8 - 9.3 - 3.1 = 6 + (-8) + (-9.3) + (-3.1)$ Add the opposites of 8, 9.3, and 3.1.

$\qquad\qquad\qquad\quad = -2 + (-9.3) + (-3.1)$ Add 6 and -8.

$\qquad\qquad\qquad\quad = -11.3 + (-3.1)$ Add -2 and -9.3.

$\qquad\qquad\qquad\quad = -14.4$ Add -11.3 and -3.1.

b. $-\frac{9}{8} + \frac{1}{4} - \left(-\frac{5}{4}\right) - \frac{1}{4} = -\frac{9}{8} + \frac{1}{4} + \left(\frac{5}{4}\right) + \left(-\frac{1}{4}\right)$ Add the opposites of $-\frac{5}{4}$ and $\frac{1}{4}$.

$\qquad\qquad\qquad\qquad = -\frac{9}{8} + \left(\frac{5}{4}\right) + \left(\frac{1}{4}\right) + \left(-\frac{1}{4}\right)$ Use commutative property.

$\qquad\qquad\qquad\qquad = -\frac{9}{8} + \frac{5}{4} + 0$ Use property of zero.

$\qquad\qquad\qquad\qquad = -\frac{9}{8} + \frac{10}{8}$ Rewrite with common denominator.

$\qquad\qquad\qquad\qquad = \frac{1}{8}$ Use rules of addition.

Evaluate the expression.

7. $\frac{11}{12} - 1 - \frac{11}{12}$ **8.** $-7 - 3.01 - (-2.2)$

9. $(-10 - 5.18) - (-10) + 5.18$

Mixed Review

Find the sum.

10. $(-13.48) + 13$ **11.** $\left(-\frac{5}{4}\right) + (-1)$

12. $634.02 + 2 + (-634.02)$ **13.** $47\frac{1}{2} + \left(-\frac{1}{2}\right) + 6 + \left(-\frac{3}{5}\right)$

14. $-\frac{1}{6} + \left(-\frac{1}{24}\right) + \left(-\frac{1}{12}\right) + \left(-\frac{1}{18}\right)$ **15.** $\$.75 + (-\$3.10) + (\$6.25) + (-\$.90)$

Topic 1

Geometry
Basic Skills Workbook: Diagnosis and Remediation

7

Quick Check

Review of Topic 1, Lesson 1

Standardized Testing Quick Check

1. Evaluate: $0.1 + 0.01 + 0.001 + 0.0001$.

A. 0.4

B. 0.0004

C. 0.1111

D. 0.0000000001

E. none of these

2. If $\dfrac{6}{13} + x = 0$, then x is equal to

A. $\dfrac{13}{6}$.

B. $\dfrac{7}{13}$.

C. $-\dfrac{13}{6}$.

D. $-\dfrac{6}{13}$.

E. none of these

Homework Review Quick Check

Find the sum.

3. $\left(-\dfrac{1}{10}\right) + \left(-\dfrac{9}{10}\right)$

4. $31.01 + (-8)$

5. $\dfrac{1}{2} + \left(-\dfrac{2}{3}\right) + (-0.5)$

Geometry
Basic Skills Workbook: Diagnosis and Remediation

Practice

For use with Lesson 1.2: Subtracting Rational Numbers

Give the opposite of each number.

1. 11.09

2. $-\dfrac{1}{4}$

3. 0

Rewrite the subtraction expression as an addition expression.

4. $-\dfrac{21}{2} - \dfrac{1}{2}$

5. $8.49 - (-0.07)$

6. $6 - 11.976$

Find the difference.

7. $6.39 - 0.18$

8. $(-6.39) - 0.18$

9. $6.39 - (-0.18)$

10. $\left(-\dfrac{5}{8}\right) - \left(-\dfrac{1}{8}\right)$

11. $\dfrac{5}{8} - \left(-\dfrac{1}{8}\right)$

12. $\left(-\dfrac{5}{8}\right) - \dfrac{1}{8}$

13. $\dfrac{11}{3} - \left(-\dfrac{11}{3}\right)$

14. $\left(-\dfrac{11}{3}\right) - \left(-\dfrac{11}{3}\right)$

15. $(-9.\overline{4}) - (-9.\overline{4})$

16. $138.97 - 2707.1$

17. $(-35) - 16.43$

18. $(-5.38) - (-219.53)$

19. $\left(-\dfrac{2}{7}\right) - \left(-\dfrac{3}{14}\right)$

20. $\left(-\dfrac{5}{8}\right) - \dfrac{2}{9}$

21. $\dfrac{5}{21} - \dfrac{2}{15}$

22. $(-167) - \dfrac{1}{4}$

23. $11\dfrac{2}{9} - \dfrac{1}{9}$

24. $\left(-153\dfrac{4}{7}\right) - \left(-1\dfrac{1}{3}\right)$

25. $36\dfrac{141}{249} - 0$

26. $0.\overline{8} - (-1)$

27. $1 - \dfrac{35}{3}$

Evaluate the expression.

28. $(-18) - 14.3 + 18$

29. $\left(-\dfrac{45}{4}\right) + \dfrac{11}{4} - \left(-\dfrac{15}{4}\right)$

30. $-72 - 1.\overline{5} - (-1.\overline{5})$

31. *Expenses*   Tammy is an art student at a local community college. She earned $635.28 in December and received a student loan of $1300. She has to pay $435.00 in tuition for school. How much money does she have left for art supplies and other living expenses?

32. *Fabric*   Mary bought 13 yards of fabric. She is cutting off two pieces to make flags. She cuts one piece $5\dfrac{1}{4}$ yards and the other piece $2\dfrac{5}{8}$ yards. How much fabric is left?

Topic 1

NAME _____ DATE _____

Multiplying and Dividing Rational Numbers

GOAL **Multiply and divide rational numbers.**

> Multiplying and dividing rational numbers are inverse operations.
> You must be able to multiply in order to divide.

Topic 1

Terms to Know	*Example/Illustration*
Reciprocal if $\dfrac{a}{b}$ is a nonzero number, then its reciprocal is $\dfrac{b}{a}$ (*Note:* Reciprocals have the same sign.)	$\dfrac{1}{4}$ and 4 $\qquad -\dfrac{3}{2}$ and $-\dfrac{2}{3}$

Understanding the Main Ideas

The rules for determining the sign of a product or quotient are the same. The box below summarizes the two rules to remember.

> **Multiplying or Dividing Signed Numbers**
>
> 1. The product or quotient of two signed numbers with the *same* sign is positive.
>
> $$-a \div -b = \frac{a}{b} \qquad\qquad \textbf{Example: } -2.2 \div -0.2 = 11$$
>
> $$-a \cdot -b = ab \qquad\qquad \textbf{Example: } -2.2 \cdot -0.2 = 0.44$$
>
> 2. The product or quotient of two signed numbers with *different* signs is negative.
>
> $$-a \div b = -\frac{a}{b} \qquad\qquad \textbf{Example: } -2.2 \div 0.2 = -11$$
>
> $$-a \cdot b = -ab \qquad\qquad \textbf{Example: } -2.2 \cdot 0.2 = -0.44$$

EXAMPLE 1 _____

Find each product or quotient.

a. $(-0.01)(-4.567)$ **b.** $8.32 \div (-0.04)$

SOLUTION

a. $(-0.01)(-4.567) = 0.04567$ Product is positive.

(continued)

NAME _____ DATE _____

Multiplying and Dividing Rational Numbers

b. $8.32 \div (-0.04) = \dfrac{8.32}{-0.04}$ Rewrite as a fraction.

$\qquad\qquad = \dfrac{8.32}{-0.04}\left(\dfrac{100}{100}\right)$ Multiply numerator and denominator by 100.

$\qquad\qquad = \dfrac{832}{-4}$ Simplify.

$\qquad\qquad = -208$ Quotient is negative.

Find the product or quotient.

  1. $4(-35.8)$ **2.** $(-6.41) \div (-0.001)$ **3.** $(-14.82)(0.03)$

You can use a reciprocal to write a division expression as a product. The division rule is particularly helpful for division expressions containing fractions.

Division Rule

To divide a number by a nonzero number b, multiply a by the reciprocal of b.

$$a \div b = a \cdot \dfrac{1}{b} \qquad\qquad \textbf{Example: } \dfrac{1}{2} \div 3 = \dfrac{1}{2} \cdot \dfrac{1}{3} = \dfrac{1}{6}$$

The result is the **quotient** of a and b.

EXAMPLE 2

Find each quotient.

  a. $\dfrac{4}{5} \div \left(-\dfrac{5}{3}\right)$ **b.** $\left(-\dfrac{1}{6}\right) \div (-3)$ **c.** $1\dfrac{1}{5} \div \left(-2\dfrac{3}{8}\right)$

SOLUTION

  a. $\dfrac{4}{5} \div \left(-\dfrac{5}{3}\right) = \dfrac{4}{5} \cdot \left(-\dfrac{3}{5}\right)$ Multiply by the reciprocal.

$\qquad\qquad = -\dfrac{12}{25}$ Product is negative.

  b. $\left(-\dfrac{1}{6}\right) \div (-3) = \left(-\dfrac{1}{6}\right) \cdot \left(-\dfrac{1}{3}\right)$ Multiply by the reciprocal.

$\qquad\qquad = \dfrac{1}{18}$ Product is positive.

  c. $1\dfrac{1}{5} \div \left(-2\dfrac{3}{8}\right) = \dfrac{6}{5} \div \left(-\dfrac{19}{8}\right)$ Rewrite as improper fractions.

$\qquad\qquad = \dfrac{6}{5} \cdot \left(-\dfrac{8}{19}\right)$ Multiply by the reciprocal.

$\qquad\qquad = -\dfrac{48}{95}$ Product is negative.

Find the quotient.

  4. $\left(-\dfrac{2}{3}\right) \div \left(-\dfrac{3}{4}\right)$ **5.** $-7 \div \left(\dfrac{7}{11}\right)$ **6.** $1\dfrac{1}{5} \div \left(-2\dfrac{1}{4}\right)$

The properties of multiplication need to be reviewed when multiplying and dividing rational numbers.

(continued)

Multiplying and Dividing Rational Numbers

Properties of Multiplication

Commutative Property

The order in which two numbers are multiplied does not change the product.

$a \cdot b = b \cdot a$ **Example:** $\frac{1}{2} \cdot \left(-\frac{1}{4}\right) = \left(-\frac{1}{4}\right) \cdot \frac{1}{2}$

Associative Property

The way you group three numbers when multiplying does not change the product.

$(a \cdot b) \cdot c = a \cdot (b \cdot c)$ **Example:** $\left[\frac{1}{2} \cdot \left(-\frac{1}{4}\right)\right] \cdot \frac{1}{3} = \frac{1}{2} \cdot \left[\left(-\frac{1}{4}\right) \cdot \frac{1}{3}\right]$

Identity Property

The product of a number and 1 is the number.

$1 \cdot a = a$ **Example:** $1 \cdot (-7.86) = -7.86$

Property of Zero

The product of a number and 0 is 0.

$a \cdot 0 = 0$ **Example:** $7.5 \cdot 0 = 0$

Property of Opposites

The product of a number and -1 is the opposite of the number.

$a \cdot (-1) = -a$ **Example:** $2.637 \cdot (-1) = -2.637$

EXAMPLE 3

Evaluate each expression.

a. $7 \cdot \frac{8}{9}\left(-\frac{1}{7}\right)$

b. $(-11.97) \div (2) \cdot (2)$

SOLUTION

a. $7 \cdot \frac{8}{9}\left(-\frac{1}{7}\right) = 7 \cdot \left(-\frac{1}{7}\right)\frac{8}{9}$ Use commutative property.

$= \left(7 \cdot -\frac{1}{7}\right)\frac{8}{9}$ Use associative property.

$= (-1)\frac{8}{9}$ Simplify.

$= -\frac{8}{9}$ Use property of opposites.

b. $(-11.97) \div (2) \cdot (2) = (-11.97) \cdot \frac{1}{2} \cdot (2)$ Rewrite as multiplication.

$= (-11.97)\left(\frac{1}{2} \cdot 2\right)$ Use associative property.

$= (-11.97)(1)$ Simplify.

$= -11.97$ Use identity property.

(continued)

NAME _____ DATE _____

Multiplying and Dividing Rational Numbers

Evaluate the expression.

7. $\dfrac{1}{12}\left(\dfrac{3}{20}\right)(-12)$

8. $-7 \div \dfrac{1}{3} \cdot \dfrac{1}{3}$

9. $(-6345.987)(899.01)(0)(21.34)$

Another property that often simplifies multiplication expressions is the distributive property.

Distributive Property

For any numbers a, b, and c,

$$a(b + c) = ab + ac$$

Example: $\dfrac{1}{2}\left(\dfrac{3}{4} + \dfrac{2}{5}\right) = \dfrac{1}{2} \cdot \dfrac{3}{4} + \dfrac{1}{2} \cdot \dfrac{2}{5}$

EXAMPLE 4

Evaluate each expression using the distributive property.

a. $(30.5)(24.2)$ **b.** $-6\left(-10\tfrac{1}{2}\right)$ **c.** $78.76(-99)$

SOLUTION

a. $(30.5)(24.2) = (30 + 0.5)(24.2)$ $30.5 = 30 + 0.5$

$\qquad\qquad\quad = (24.2)(30 + 0.5)$ Use commutative property.

$\qquad\qquad\quad = (24.2)(30) + (24.2)(0.5)$ Use distributive property.

$\qquad\qquad\quad = 726 + 12.1$ Simplify.

$\qquad\qquad\quad = 738.1$ Simplify.

b. $-6\left(-10\tfrac{1}{2}\right) = -6\left[-10 + \left(-\tfrac{1}{2}\right)\right]$ $-10\tfrac{1}{2} = -10 + \left(-\tfrac{1}{2}\right)$

$\qquad\qquad\quad = 60 + 3$ Use distributive property.

$\qquad\qquad\quad = 63$ Simplify.

c. $78.76(-99) = 78.76(-100 + 1)$ $-99 = -100 + 1$

$\qquad\qquad\quad = -7876 + 78.76$ Use distributive property.

$\qquad\qquad\quad = -7797.24$ Simplify.

Evaluate the expression using the distributive property.

10. $(12.8)(10.5)$ **11.** $15\left(-2\tfrac{1}{5}\right)$ **12.** $43.26(-98)$

Mixed Review

Evaluate the expression.

13. $(-4.93) - (-1)$ **14.** $\left(\dfrac{16}{3}\right) + \left(-\dfrac{3}{4}\right) - \left(\dfrac{1}{3}\right)$ **15.** $-36.11 - 0.062 - 2.9$

NAME _____ DATE _____

Quick Check

Review of Topic 1, Lesson 2

Standardized Testing Quick Check

1. Which of the following has the same value as $-8 - (-3.2)$?

   A. $-8 + (-3.2)$

   B. $8 + (-3.2)$

   C. $-8 + (3.2)$

   D. $8 + 3.2$

   E. none of these

2. What is the opposite of $-\frac{1}{3}$?

   A. -3

   B. 3

   C. $-\frac{2}{3}$

   D. $\frac{1}{3}$

   E. none of these

Homework Review Quick Check

Find the difference.

3. $(-19) - \left(-\frac{7}{8}\right)$

4. $63.547 - 3.093$

5. $\frac{1}{3} - \left(-\frac{29}{6}\right) - (0.\overline{3})$

NAME _____ DATE _____

Practice

For use with Lesson 1.3: Multiplying and Dividing Rational Numbers

Find the reciprocal of the number.

1. -81

2. $\dfrac{6}{7}$

3. $5\dfrac{3}{10}$

Rewrite the division expression as a multiplication expression.

4. $\dfrac{7}{11} \div (-4)$

5. $\left(-\dfrac{1}{9}\right) \div \left(-\dfrac{1}{8}\right)$

6. $34\dfrac{1}{2} \div 1\dfrac{1}{2}$

Find the product or quotient.

7. $(15.9)(-0.0001)$

8. $(-1.896) \div (-10)$

9. $0.1 \div (-0.1)$

10. $\left(-\dfrac{2}{3}\right)\left(-\dfrac{1}{3}\right)$

11. $\left(-\dfrac{2}{3}\right) \div \left(-\dfrac{1}{3}\right)$

12. $\left(-\dfrac{2}{3}\right)(-3)$

13. $\left(\dfrac{5}{4}\right)\left(-\dfrac{3}{10}\right)$

14. $\left(-\dfrac{7}{5}\right)\left(-\dfrac{7}{5}\right)$

15. $\left(-\dfrac{7}{5}\right) \div \left(-\dfrac{7}{5}\right)$

16. $(64.12)(5.5)$

17. $(-125.5) \div 0.5$

18. $(-39.03) \div (-0.3)$

19. $\left(-1\dfrac{3}{4}\right) \div \left(-2\dfrac{1}{4}\right)$

20. $\left(-\dfrac{5}{6}\right)\left(6\dfrac{2}{18}\right)$

21. $\left(56\dfrac{3}{13}\right)(0)$

22. $(-16)\left(\dfrac{1}{4}\right)$

23. $(-16) \div \dfrac{1}{4}$

24. $(-16) \div (-4)$

25. $(-98)(272.63)$

26. $(7777)\left(-1\dfrac{1}{7}\right)$

27. $4848 \div 48.48$

Evaluate the expression.

28. $\left(-\dfrac{4}{9}\right)(-1) \div \dfrac{4}{9}$

29. $\left(-\dfrac{42}{9}\right)\left(\dfrac{12}{7}\right)\left(-\dfrac{15}{6}\right)$

30. $9 \div \dfrac{2}{9} \cdot \left(-\dfrac{2}{9}\right)$

31. *Pencils* Eugena is buying 4 packs of pencils that cost $1.25 each. If the sales tax is $0.06 per dollar, how much is the sales tax on the 4 packs of pencils?

32. *Pizza* Ray has 4 pizzas and plans to give $\frac{1}{5}$ of a pizza to each of his friends. How many slices can he give away?

Topic 1

Order of Operations

GOAL Use the order of operations to evaluate numerical expressions.

> One way to avoid confusion when communicating algebraic
> concepts is to use the established order of operations.

Understanding the Main Ideas

Numerical expressions may contain grouping symbols and more than one operation. An agreement called the order of operations sets the priorities to use when
evaluating these expressions. Without the order of operations, there would be
inconsistency and disagreement concerning the values of numerical expressions.

> **Order of Operations**
>
> To evaluate an expression involving more than one operation, use the
> following order.
> **1.** First do operations that occur within grouping symbols.
> **2.** Then evaluate powers.
> **3.** Then do multiplications and divisions from left to right.
> **4.** Finally do additions and subtractions from left to right.

If there are no grouping symbols, the order of operations begins with evaluating
powers.

EXAMPLE 1 _____

Evaluate each expression.

a. $4 + 2\left(\frac{1}{2}\right)^2$

b. $-7.2 - (4)^2 \div (-0.5)$

SOLUTION

a. $4 + 2\left(\frac{1}{2}\right)^2 = 4 + 2\left(\frac{1}{4}\right)$　　　　　　　Evaluate power.

$\qquad\qquad\quad = 4 + \left(\frac{1}{2}\right)$　　　　　　　Evaluate product.

$\qquad\qquad\quad = 4\frac{1}{2}$　　　　　　　　　Evaluate sum.

b. $-7.2 - (4)^2 \div (-0.5) = -7.2 - (16) \div (-0.5)$　　　Evaluate power.

$\qquad\qquad\qquad\qquad = -7.2 - (-32)$　　　Evaluate quotient.

$\qquad\qquad\qquad\qquad = 24.8$　　　　　　　Evaluate difference.

Evaluate the expression.

1. $-1 + \left(\frac{1}{3}\right)^2 \div \frac{1}{18}$

2. $3(0.01)^3 - 2.41$

3. $5 \cdot (6.9) + (1.1)^2$

(continued)

NAME _____   DATE _____

Order of Operations

Multiplication and division have priority over addition and subtraction; yet
they have the same priority with each other as do addition and subtraction.
In expressions with both multiplication and division *or* addition and subtraction,
the left-to-right rule is used.

EXAMPLE 2

Evaluate each expression.

a. $\frac{1}{4} \div \frac{3}{4} \cdot (-3)$ **b.** $9.21 - 0.21 + 1$ **c.** $8.5 \div 0.01 \cdot (-2) - 0.02 + (75)(0.6)$

SOLUTION

a. $\frac{1}{4} \div \frac{3}{4} \cdot (-3) = \left(\frac{1}{4} \div \frac{3}{4}\right) \cdot (-3)$ Work from left to right.

$\phantom{\frac{1}{4} \div \frac{3}{4} \cdot (-3)} = \left(\frac{1}{4} \cdot \frac{4}{3}\right) \cdot (-3)$ Multiply by the reciprocal.

$\phantom{\frac{1}{4} \div \frac{3}{4} \cdot (-3)} = \left(\frac{1}{3}\right) \cdot (-3)$ Simplify.

$\phantom{\frac{1}{4} \div \frac{3}{4} \cdot (-3)} = -1$ Multiply.

b. $9.21 - 0.21 + 1 = (9.21 - 0.21) + 1$ Work from left to right.

$ = 9.00 + 1$ Subtract.

$ = 10$ Add.

c. $8.5 \div 0.01 \cdot (-2) - 0.02 + (75)(0.6) = (850)(-2) - 0.02 + 45$ Work from left to right.

$ = -1700 - 0.02 + 45$ Multiply.

$ = -1700.02 + 45$ Subtract.

$ = -1655.02$ Add.

Evaluate the expression.

4. $156.3 \div 0.3 \cdot (-5)$ **5.** $\frac{17}{8} - \frac{1}{8} + 13$

6. $(-3) \div 0.1 \cdot 0.5 - 8 + 4.3 \cdot 0.5$

Placing grouping symbols into a numerical expression is the way to indicate that
addition is to be done before multiplication or multiplication before exponents.
Parentheses are the grouping symbols that are used to override the stated operation priorities. The first step in the order of operations is to perform operations
within grouping symbols. Note that brackets, [ ], are used for the same purpose
as ( ) except in the calculator where they have a different meaning.

(continued)

Order of Operations

EXAMPLE 3

Evaluate each expression.

a. $-21(4.6 - 4.5)^2$ **b.** $\left(\frac{5}{6} - \frac{1}{6}\right)\left(\frac{3}{4} + \frac{1}{4}\right)$ **c.** $(1.2)(2^3 + 4 \div 2)$

SOLUTION

a. $-21(4.6 - 4.5)^2 = -21(0.1)^2$ Subtract inside the parentheses.

$ = -21(0.01)$ Evaluate power.

$ = -0.21$ Multiply.

b. $\left(\frac{5}{6} - \frac{1}{6}\right)\left(\frac{3}{4} + \frac{1}{4}\right) = \left(\frac{4}{6}\right)\left(\frac{4}{4}\right)$ Evaluate inside parentheses.

$\phantom{\left(\frac{5}{6} - \frac{1}{6}\right)\left(\frac{3}{4} + \frac{1}{4}\right)} = \frac{4}{6}$ Multiply.

$\phantom{\left(\frac{5}{6} - \frac{1}{6}\right)\left(\frac{3}{4} + \frac{1}{4}\right)} = \frac{2}{3}$ Simplify.

c. $(1.2)(2^3 + 4 \div 2) = (1.2)(8 + 4 \div 2)$ Evaluate power inside parentheses.

$ = (1.2)(8 + 2)$ Divide inside parentheses.

$ = (1.2)(10)$ Add inside parentheses.

$ = 12$ Multiply.

Evaluate the expression.

7. $(3.28 + 14.51) \div (2.2 - 2.1)$ **8.** $-80\left(\frac{3}{2} + \frac{5}{2}\right)^2$ **9.** $231 - (4.2 - 2^4 \div 4)$

Absolute value symbols and fraction bars are implied grouping symbols. All operations inside absolute value symbols must be completed in order before the absolute value is applied. All operations in the numerator or denominator of a fraction must be performed before the division of the fraction bar is applied.

EXAMPLE 4

Evaluate each expression.

a. $\left| -\frac{4}{3} + \frac{1}{6} \right|$ **b.** $-11|8.3 - (-2.7)| + 7.63$ **c.** $\dfrac{5^2 - 10(0.4)}{1 - 2^3}$

SOLUTION

a. $\left| -\frac{4}{3} + \frac{1}{6} \right| = \left| -\frac{8}{6} + \frac{1}{6} \right|$ Rewrite using the LCD.

$\phantom{\left| -\frac{4}{3} + \frac{1}{6} \right|} = \left| -\frac{7}{6} \right|$ Add inside the absolute value symbols.

$\phantom{\left| -\frac{4}{3} + \frac{1}{6} \right|} = \frac{7}{6}$ Take absolute value.

(continued)

Order of Operations

b. $-11|8.3 - (-2.7)| + 7.63 = -11|11| + 7.63$ Subtract inside absolute value symbols.

$= -11(11) + 7.63$ Take absolute value.

$= -121 + 7.63$ Multiply.

$= -113.37$ Add.

c. $\dfrac{5^2 - 10(0.4)}{1 - 2^3} = \dfrac{25 - 10(0.4)}{1 - 8}$ Evaluate powers.

$= \dfrac{25 - 4}{-7}$ Multiply in numerator and subtract in denominator.

$= \dfrac{21}{-7}$ Subtract in numerator.

$= -3$ Divide.

Evaluate the expression.

10. $|-137.54 + 67.38|$

11. $9\left|-6 \div \dfrac{1}{3} + 4^3\right|$

12. $\dfrac{14 \div 7 + 8 \cdot (-4)}{-9 - 1^8}$

Mixed Review

Evaluate the expression.

13. $-4 \div \dfrac{8}{13}$

14. $-125 - (-5.25)$

15. $\left(6\dfrac{2}{5}\right)(7)$

Quick Check

Review of Topic 1, Lesson 3

Standardized Testing Quick Check

1. The reciprocal of $-\dfrac{7}{8}$ is

   A. $-\dfrac{8}{7}$.
                   B. $\dfrac{7}{8}$.

   C. $\dfrac{8}{7}$.
                   D. $\dfrac{1}{8}$.

   E. none of these

2. Which expression is equivalent to $13(-96)$?

   A. $13(100-4)$
                B. $13\left(-\dfrac{1}{96}\right)$

   C. $13(-100+4)$
             D. $\dfrac{1}{13}(-96)$

   E. none of these

Homework Review Quick Check

Evaluate the expression.

3. $\dfrac{5}{6} \div (-5)$
          4. $(1.346)(1.1)$
          5. $\dfrac{4}{9} \div 4 \cdot (-9)$

Practice

For use with Lesson 1.4: Order of Operations

Evaluate the expression.

1. $-8.1 + 4(-5)$

2. $(-8.1 + 4)(-5)$

3. $6 \cdot \left(\dfrac{1}{2}\right) \div 8 \cdot (8)$

4. $6\left(\dfrac{1}{2}\right) + 8(8)$

5. $6\left[\dfrac{1}{2} \div 8 \cdot (8)\right]$

6. $\left(6 \cdot \dfrac{1}{2}\right) \div (8 \cdot 8)$

7. $(0.4)^3 - (0.5)(30)$

8. $[(0.4)^3 - (0.5)]30$

9. $(0.5)(30) - (0.4)^3$

10. $\left(-\dfrac{2}{7}\right)^2\left(-\dfrac{7}{10}\right) + 1$

11. $\left(-\dfrac{2}{7}\right)^2\left(-\dfrac{7}{10} + 1\right)$

12. $\left(-\dfrac{2}{7} - \dfrac{7}{10}\right)^2 + 1$

13. $\dfrac{10}{3} \div \left(-\dfrac{3}{10}\right) \cdot 9$

14. $\dfrac{10}{3} \div \left(-\dfrac{3}{10} \cdot 9\right)$

15. $\left(\dfrac{10}{3}\right) \div \left|-\dfrac{3}{10} \cdot 9\right|$

16. $1.2 + (0.1)^4 - 3$

17. $7 - 3\left(\dfrac{1}{4}\right)^2$

18. $(7 - 3)\left(\dfrac{1}{4}\right)^2$

19. $19 - (3.2 - 0.2)$

20. $19 - 3.2 - 0.2$

21. $(19 - 3.2) - 0.2$

22. $54.2 \div 0.2 \cdot (-6)$

23. $\dfrac{23}{11} - \dfrac{1}{11} + 67$

24. $(-24) \div 0.2 \cdot 0.4 - 21 + (7.4)(0.01)$

25. $\left(3\dfrac{1}{2} + 16\dfrac{1}{2}\right) \div \left(\dfrac{1}{3} - 2\dfrac{1}{3}\right)$

26. $-\dfrac{3}{4}(5 - 6)^2$

27. $81.95 - [1 - (1.2)^2 \div 2]$

28. $|-3593.2 - 1200.5|$

29. $-100\left|7 \div \left(-\dfrac{21}{5}\right) + 1^6\right|$

30. $\dfrac{-65 \div 13 + 91 \cdot (-3)^2}{2^3 - 12 \div 3}$

31. *Restaurant Bill*   Most restaurants add a 15% gratuity charge to the bills
of diners who are in groups of six or more. You and eight friends have dinner
in such a restaurant. Your bill is $8.95 before taxes or gratuity. The tax is 6%.
What is your total bill? (Note: the gratuity is determined using the bill with
tax added.)

NAME _____ DATE _____

Assessment

For use with Topic 1: Operations with Rational Numbers

Find the absolute value.

1. $|-8.5|$

2. $\left|\dfrac{1}{3}\right|$

3. $\left|-4\dfrac{3}{8}\right|$

Give the opposite of each number.

4. 1

5. -75

6. 25

Find the sum or difference.

7. $\dfrac{5}{6} + \dfrac{1}{6}$

8. $\left(-\dfrac{2}{5}\right) + \left(-\dfrac{2}{3}\right)$

9. $0.56 - 8.87$

10. $10.25 + (-14.33)$

11. $1 + \left(-\dfrac{13}{15}\right)$

12. $(-2.56) - (-2.56)$

13. $\dfrac{9}{10} - \dfrac{3}{5}$

14. $\left(-\dfrac{8}{3}\right) - \left(-\dfrac{4}{7}\right)$

15. $-3.65 + (-19.01)$

Find the reciprocal of the number.

16. -23

17. $\dfrac{6}{7}$

18. 0.25

Find the product or quotient.

19. $\dfrac{4}{9} \div \dfrac{8}{3}$

20. $(-2.3)(8.4)$

21. $\dfrac{3}{11} \cdot \left(-\dfrac{5}{18}\right)$

22. $(-65.5) \div (-0.5)$

23. $\left(-3\dfrac{2}{3}\right) \div \left(-1\dfrac{3}{6}\right)$

24. $14.25 \div 3.8$

25. $\left(-\dfrac{16}{5}\right) \cdot \left(\dfrac{25}{28}\right)$

26. $(10.8)(-6.57)$

27. $\left(-\dfrac{9}{8}\right) \div \left(-3\dfrac{3}{4}\right)$

Evaluate the expression.

28. $\dfrac{4}{5} \div \left(-\dfrac{1}{5}\right) + 13$

29. $\left|\dfrac{9}{10} \cdot \dfrac{1}{2}\right| - \dfrac{3}{10}$

30. $3(2.7 \div 0.9) - 5$

31. $\dfrac{1}{2} \cdot 26 - 3^2$

32. $2.5 \cdot (-0.5)^2 \div 5$

33. $\dfrac{1}{3}(9 \cdot 3) + 18$

34. $\dfrac{9 \cdot 2}{4 + 3^2 - 1}$

35. $\dfrac{13 - (-4)}{18 - 4^2 + 1}$

36. $\dfrac{5^3 \cdot 2}{1 + 6^2 - 8}$

NAME _____ DATE _____

Evaluating Expressions

GOAL **Evaluate algebraic expressions.**

Variables are used in algebraic expressions to express number relationships. For example, $A = lw$ is the area of a rectangle where l is length and w is width. If $l = 3$ ft and $w = 2$ ft, we can evaluate the expression and find that the area is $3 \cdot 2 = 6$ ft^2.

Terms to Know   *Example/Illustration*

Terms to Know	Example/Illustration
Variable a letter that is used to represent one or more numbers	x, y, a, b, A, α
Algebraic expression a collection of numbers, variables, operations, and grouping symbols (Expressions **do not** contain connecting symbols such as $=$, $<$, or $>$.)	$3x \quad 4y - 1 \quad a^2 \quad 5(b - 8)$ $bh \quad \|A - 3B\|$
Value of the variable the numbers that the variable represents	Evaluate $3x$ when $x = -4$; -4 is the **value** of x.
Value of the expression the numerical value of the algebraic expression after the variables have been replaced by their individual values	The value of $3x$ when $x = -4$ is $3(-4)$ or -12.

Understanding the Main Ideas

When each variable in an algebraic expression is replaced by a number, we say that we are *evaluating* the expression, and the resulting number is the value of the expression.

EXAMPLE 1

Evaluate each expression when $a = -3$, $b = 0.1$, and $c = \frac{1}{3}$.

a. $4 - (a - 3)$ **b.** $5a + b$ **c.** $\dfrac{c}{a}$

SOLUTION

a. $4 - (a - 3) = 4 - (-3 - 3)$ Substitute -3 for a.

               $= 4 - (-6)$ Evaluate inside parentheses.

               $= 10$ Simplify.

(continued)

Evaluating Expressions

b. $5a + b = 5(-3) + 0.1$ Substitute -3 for a and 0.1 for b.

$\qquad = -15 + 0.1$ Multiply.

$\qquad = -14.9$ Simplify.

c. $\dfrac{c}{a} = \dfrac{\left(\frac{1}{3}\right)}{-3}$ Substitute -3 for a and $\frac{1}{3}$ for c.

$\qquad = \dfrac{1}{3} \div -3$ Fraction bar indicates division.

$\qquad = \dfrac{1}{3} \cdot -\dfrac{1}{3}$ Rewrite as multiplication.

$\qquad = -\dfrac{1}{9}$ Simplify.

Evaluate the expression when $x = -\frac{1}{5}$, $y = 6$, and $z = 0.2$.

1. $5x - y$ **2.** $\dfrac{y}{z}$ **3.** $-(xy) + \dfrac{1}{10}$

When evaluating expressions that contain exponents, remember that the exponent applies only to the variable or number immediately in front of it unless parentheses are used.

EXAMPLE 2

Evaluate each expression when $x = \frac{2}{3}$ and $y = -1$.

a. $-y^2$ **b.** $(3x)^3$ **c.** $3 - (x + y)^2$

SOLUTION

a. $-y^2 = -(-1)^2$ Substitute -1 for y.

$\qquad = -(1)$ Evaluate power.

$\qquad = -1$ Simplify.

b. $(3x)^3 = \left(3 \cdot \frac{2}{3}\right)^3$ Substitute $\frac{2}{3}$ for x.

$\qquad = (2)^3$ Multiply within parentheses.

$\qquad = 8$ Evaluate power.

c. $3 - (x + y)^2 = 3 - \left[\frac{2}{3} + (-1)\right]^2$ Substitute $\frac{2}{3}$ for x and -1 for y.

$\qquad = 3 - \left(-\frac{1}{3}\right)^2$ Add within parentheses.

$\qquad = 3 - \left(\frac{1}{9}\right)$ Evaluate power.

$\qquad = 2\frac{8}{9}$ Subtract.

(continued)

Topic 2

Evaluating Expressions

Evaluate the expression when $a = 2$ and $b = -\frac{3}{4}$.

4. $-2b^2$ **5.** $(-a)^3 + 7$ **6.** $6 + 2(a + b)^2$

Remember that absolute value symbols and fraction bars are implied grouping symbols.

EXAMPLE 3

Evaluate the expression when $x = 4.1$, $y = -0.4$, and $z = 10$.

a. $|x + y|$ **b.** $\dfrac{xz - y}{z^3}$

SOLUTION

a. $|x + y| = |4.1 + (-0.4)|$ Substitute 4.1 for x and -0.4 for y.

$\qquad\qquad = |3.7|$ Add within absolute value symbols.

$\qquad\qquad = 3.7$ Simplify.

b. $\dfrac{xz - y}{z^3} = \dfrac{4.1 \cdot 10 - (-0.4)}{10^3}$ Substitute 4.1 for x, -0.4 for y, and 10 for z.

$\qquad\qquad = \dfrac{41 + 0.4}{1000}$ Simplify numerator and denominator.

$\qquad\qquad = \dfrac{41.4}{1000}$ Simplify numerator.

$\qquad\qquad = 0.0414$ Divide.

Evaluate the expression when $c = 5.7$ and $d = -1.3$.

7. $|c - d|$ **8.** $\dfrac{3d^2 - c}{c + d}$ **9.** $\dfrac{c - d}{|d| + 0.7}$

Formulas contain algebraic expressions that represent real-life relationships. Some important formulas for area and perimeter of rectangles, squares, triangles, and circles are listed below and on the next page.

Perimeter and Area Formulas

Rectangle

Area $= lw$, where l is the length and w is the width

Perimeter $= 2l + 2w$

Square

Area $= s^2$, where s is the side length

Perimeter $= 4s$

(continued)

Topic 2

NAME _____ DATE _____

Evaluating Expressions

Triangle

Area $= \frac{1}{2}bh$, where b is the base and h is the height

Perimeter $= a + b + c$, where a, b, and c are the side lengths of the triangle

Circle

Area $= \pi r^2$, where r is the radius and $\pi \approx 3.14$

Circumference $= 2\pi r$

Using a formula requires evaluating algebraic expressions.

EXAMPLE 4

Find the perimeter of a rectangle with length 3.2 cm and width 1.6 cm.

SOLUTION

Perimeter $= 2l + 2w$	Write formula.
$= 2(3.2) + 2(1.6)$	Substitute 3.2 for l and 1.6 for w.
$= 6.4 + 3.2$	Simplify.
$= 9.6$ cm	Write solution with units.

10. Find the area of a triangle with $b = 4$ in. and $h = \frac{3}{2}$ in.

11. Find the circumference of a circle with radius 2.3 meters.

12. Find the area of a square with side length 10 feet.

Mixed Review

Evaluate the expression.

13. $\left(\dfrac{5}{6}\right)\left(\dfrac{3}{25}\right) - 10 \div \dfrac{1}{5}$ **14.** $-(1.5)^2 + 4$ **15.** $\dfrac{12(0.4) + 0.2}{|-1 - 4|}$

NAME _____ DATE _____

Quick Check

Review of Topic 1, Lesson 4

Standardized Testing Quick Check

1. What is the first operation used in simplifying
   $5 - 3 \cdot (6.7) \div (0.1) + (2.1)$?

   A. addition

   B. subtraction

   C. multiplication

   D. division

   E. none of these

2. Which of the following is *not* an implied grouping symbol?

   A. absolute value symbol

   B. fraction bar

   C. parentheses

   D. percent sign

   E. none of these

Homework Review Quick Check

Evaluate the expression.

3. $31 \div 10 \cdot 0.4$

4. $\dfrac{1}{6} \cdot 5 + \left(-\dfrac{1}{12}\right)$

5. $9 - (7.86 - 2.04)$

Topic 2

Practice

For use with Lesson 2.1: Evaluating Expressions

Evaluate the expression when $x = -6$ and $y = \frac{2}{3}$.

1. xy

2. $3y$

3. $-6y - x$

4. $\frac{1}{6}x$

5. $\frac{y}{2}$

6. $-|x|$

7. $x^2 + y^2$

8. $-3xy$

9. $17 + 2(x - 1)$

10. $y + 3$

11. $(x + 1)^2$

12. $|x - 3| + 12y$

13. $-x^2 + 100$

14. $|x| + |y|$

15. $-21y \div 3$

Evaluate the expression when $a = 0.07$, $b = -1$, and $c = 100$.

16. $a + b + c$

17. $ab - ac$

18. $a(b - c)$

19. $|-a|$

20. $c - b - a$

21. $c - (b - a)$

22. $b + \frac{a}{c}$

23. $\frac{b + a}{c}$

24. b^{20}

25. $15 - c|a + b|$

26. $a^2 + b^2 + c^2$

27. $4a + 3b$

28. $7(a + b)$

29. $\frac{ac}{b^3}$

30. $ac - b + a$

31. Find the perimeter of a rectangle with $l = 7.4$ meters and $w = 3.8$ meters.

32. Find the area of a circle with a diameter of 3 inches.

33. Find the perimeter of a square with side equal to $3\frac{2}{5}$ centimeters.

34. Use the formula for distance, $d = rt$, to find the distance d traveled when a car travels at the constant rate r of 75 kilometers per hour for one half of an hour. (t is time in hours.)

35. Use the formula $S = 6s^2$ to find the surface area S of a cube with side lengths of 4 feet.

Topic 2

NAME _____ DATE _____

Simplifying Linear Expressions in One Variable

GOAL Simplify linear expressions in one variable.

> The distributive property allows you to simplify linear expressions by combining like terms.

Terms to Know

Example/Illustration

Terms to Know	Example/Illustration
Linear expression any expression of the form $ax + b$, where a and b are real numbers and x is any variable (a is the coefficient of x, and a or b could equal 0.)	$3x + 2$ $5y$ 43 $\frac{1}{2}a - 62$
Terms in the linear expression $ax + b$, ax and b are the terms	**Expression** **Terms** $3x + 2$ $3x$ and 2 (2 terms) $5y$ $5y$ (one term) 43 43 (one **constant** term) $\frac{1}{2}a - 62$ $\frac{1}{2}a$ and -62 (2 terms)
Like terms in two or more linear expressions, all terms of the form ax are called like terms when they contain the same variable (All constant terms b are like terms.)	In expressions $3x + 2$ and $7x - 5$, the like terms are $3x$ and $7x$; 2 and -5 are the constant terms. In the expressions $-6y + 3$ and $5y$, the like terms are $-6y$ and $5y$.

Topic 2

Understanding the Main Ideas

The sums and differences of linear expressions can be combined into simpler expressions by using the distributive property.

> **Distributive Property**
> For any numbers a, b, and c,
> $$a(b + c) = ab + ac \quad \text{and} \quad ab + ac = a(b + c)$$
> $$a(b - c) = ab - ac \quad \text{and} \quad ab - ac = a(b - c)$$

EXAMPLE 1

Use the distributive property to simplify each linear expression.

a. $2(x + 4) + (-9x - 1)$ **b.** $7.3a - 0.8 + 3.7a + 1.2$

(continued)

NAME _____   DATE _____

Simplifying Linear Expressions in One Variable

SOLUTION

a. $2(x + 4) + (-9x - 1) = 2x + 8 + (-9x - 1)$ Distribute the 2.

$= 2x - 9x + 8 - 1$ Group like terms.

$= (2 - 9)x + 8 - 1$ Use distributive property.

$= -7x + 7$ Simplify.

b. $7.3a - 0.8 + 3.7a + 1.2 = 7.3a + 3.7a - 0.8 + 1.2$ Group like terms.

$= (7.3 + 3.7)a - 0.8 + 1.2$ Use distributive property.

$= 11a + 0.4$ Simplify.

Simplify.

1. $(12b + 1.3) + (-b - 0.7)$ **2.** $-\dfrac{1}{2}x + \dfrac{1}{2}x$ **3.** $4.3(c + 2) - 5.7 + c$

Differences of linear expressions can also be simplified.

EXAMPLE 2

Topic 2

Use the distributive property to simplify each difference of linear expressions.

a. $\frac{5}{6}d - \frac{2}{3}d$ **b.** $3 - 2(4 + x)$

SOLUTION

a. $\frac{5}{6}d - \frac{2}{3}d = \left(\frac{5}{6} - \frac{2}{3}\right)d$ Use distributive property.

$= \left(\frac{5}{6} - \frac{4}{6}\right)d$ The least common denominator is 6.

$= \frac{1}{6}d$ Simplify.

b. $3 - 2(4 + x) = 3 + (-2)(4 + x)$ Rewrite as an addition expression.

$= 3 + [(-2)(4) + (-2)(x)]$ Distribute the -2.

$= 3 + (-8) + (-2x)$ Multiply.

$= -5 + (-2x)$ Combine like terms.

$= -5 - 2x$ Simplify.

Simplify.

4. $1.3x - 2.1x$ **5.** $7x - 3(x - 4)$ **6.** $-4(x + 2) - 6x$

Mixed Review

7. Evaluate $|x^3 - y|$ when $x = -3$ and $y = 7$.

8. Find the perimeter of a triangle with side lengths of 4.56 inches, 3.96 inches, and 2.13 inches.

9. Evaluate: $-5(3.7 + 1.3) + 4(8.5 - 0.5)$.

10. Evaluate: $89 - 4(-5)^2$.

NAME _____ DATE _____

Quick Check

Review of Topic 2, Lesson 1

Standardized Testing Quick Check

1. Evaluate $-m^4$ when $m = -1$.

   A. -1

   B. 1

   C. -4

   D. 4

   E. none of these

2. Which of the following is equivalent to the expression $|3 - 4|$?

   A. $|3| + |-4|$

   B. $|3| - |4|$

   C. $|-1|$

   D. 1 or -1

   E. none of these

Homework Review Quick Check

Evaluate the expression when $a = \frac{9}{8}$, $b = -16$, **and** $c = 1.5$.

3. $ab - 4c$

4. $(2c - b)^2$

5. $\dfrac{a}{|bc|}$

Topic 2

NAME _____ DATE _____

Practice

For use with Lesson 2.2: Simplifying Linear Expressions in One Variable

Decide if each of the following is a linear expression.

1. $x^2 + 5$ **2.** $-7y - 3$ **3.** $11a$

4. $18 + 46b$ **5.** $|x| + 9$ **6.** $4xy$

Simplify by combining like terms.

7. $14x + x$ **8.** $-y - y$ **9.** $5a + 6a + 17$

10. $-3c - 3 + 10c - 7$ **11.** $15 + d + 6d$ **12.** $87h - 21h$

13. $13b + 9 - 2b - 9$ **14.** $(3y + 1)(-2) + y$ **15.** $2(x + 1) - 3(x - 4)$

16. $(5.1x + 0.06) + (x - 1)$ **17.** $4(2 - a) - a$ **18.** $2359 + 145 + 21x$

19. $\dfrac{5}{6}k + \dfrac{1}{6}k$ **20.** $3n - 7 + \dfrac{1}{2}n + 8$ **21.** $60 - \dfrac{6}{7}y + \dfrac{8}{7}y$

22. $12x + (7 - x)2$ **23.** $-91.26z + 2(2.51z + 3.22)$ **24.** $12(2n - 3) + 4(n - 13)$

25. $\dfrac{5}{14}y + y + \dfrac{1}{7}y$ **26.** $-\dfrac{19}{8}v - \dfrac{5}{24}v - \dfrac{8}{3}v$ **27.** $6\dfrac{1}{4} + 5c - \dfrac{3}{8}$

28. $(7 - 6.43x) - 7x$ **29.** $8d - 900 - d + 45.7$ **30.** $-16t + 21 + (-0.08t)$

31. The length of the rectangle below is twice the width. Find a simplified algebraic expression for the perimeter of the rectangle.

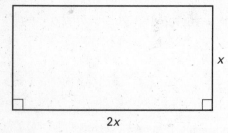

32. Find a simplified algebraic expression for the perimeter of the quadrilateral below.

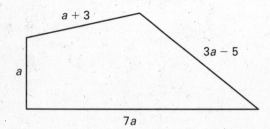

Topic 2

NAME _____ DATE _____

Simplifying Expressions in Two Variables

GOAL Simplify algebraic expressions in two variables.

> Algebraic expressions in two variables can also be simplified using the distributive property.

Terms to Know	Example/Illustration	
Like terms terms that contain the same variable part (The coefficients may differ.)	**Expression**	**Like Terms**
	$3x + 5y$ and $4x - 6y$	$3x$ and $4x$ $5y$ and $-6y$
	$7x^2 - 4xy$ and $x^2 + 8xy$	$7x^2$ and x^2 $-4xy$ and $8xy$

Understanding the Main Ideas

The distributive property may also be applied to each set of like terms in an algebraic expression to combine them into a simpler expression.

EXAMPLE 1
_____

Simplify by combining like terms.

a. $x + y + 2x + y$ **b.** $7y^2 - y^2$ **c.** $13ab - 9 + 5ab + 8$

SOLUTION

a. $x + y + 2x + y = x + 2x + y + y$ Group like terms.

              $= (1 + 2)x + (1 + 1)y$ Use distributive property.

              $= (3)x + (2)y$ Add inside parentheses.

              $= 3x + 2y$ Simplify.

b. $7y^2 - y^2 = (7 - 1)y^2$ Use distributive property.

            $= (6)y^2$ Subtract inside parentheses.

            $= 6y^2$ Simplify.

c. $13ab - 9 + 5ab + 8 = 13ab + 5ab - 9 + 8$ Group like terms.

                      $= (13 + 5)ab - 9 + 8$ Use distributive property.

                      $= 18ab - 1$ Add inside parentheses and add constants.

Simplify by combining like terms.

1. $(2a + b) + (-10a + 3b)$ **2.** $71x^2 + 4x^2$ **3.** $15cd + cd + 7 - 9$

(continued)

Topic 2

NAME _____ DATE _____

Simplifying Expressions in Two Variables

Terms are alike if the variable portions of the terms including the exponents are exactly alike. When combining like terms with exponents, add the coefficients but do *not* change the exponents.

EXAMPLE 2

Combine like terms.

a. $11x^2 + 4x + 7 - 3x^2 - x$

b. $(4x^2 - 5xy + y) - (7x^2 + xy - 6y)$

c. $-c^3d + 4c^2d + 4c^3d$

SOLUTION

a. $11x^2 + 4x + 7 - 3x^2 - x = 11x^2 - 3x^2 + 4x - x + 7$ Group like terms.

$= (11 - 3)x^2 + (4 - 1)x + 7$ Use distributive property.

$= 8x^2 + 3x + 7$ Simplify.

b. $(4x^2 - 5xy + y) - (7x^2 + xy - 6y)$

$= 4x^2 - 5xy + y - 7x^2 - xy + 6y$ Use distributive property.

$= 4x^2 - 7x^2 - 5xy - xy + y + 6y$ Group like terms.

$= (4 - 7)x^2 + (-5 - 1)xy + (1 + 6)y$ Use distributive property.

$= -3x^2 - 6xy + 7y$ Simplify.

c. $-c^3d + 4c^2d + 4c^3d = -c^3d + 4c^3d + 4c^2d$ Group like terms.

$= (-1 + 4)c^3d + 4c^2d$ Use distributive property.

$= 3c^3d + 4c^2d$ Add inside parentheses.

Simplify by combining like terms.

4. $31s^2 - 17s + 8 + 9s^2 - 6s - 5$ **5.** $(a^2 + ab + b^2) + (4ab + b^2)$

6. $3(x - y) - (2x + y)$

Mixed Review

7. Simplify: $12.3z + 0.07z - 1.99z$.

8. Find the perimeter of a square with side length $5x - 7$.

9. Evaluate $-\left(\dfrac{3x}{4}\right)^2$ when $x = -4$.

10. Evaluate: $-96(153.72)$.

NAME _____ DATE _____

Quick Check

Review of Topic 2, Lesson 2

Standardized Testing Quick Check

1. Which expression is equivalent to $-11x + 4x$?

    A. $-7x^2$

    B. $-7x$

    C. $7x$

    D. $7x^2$

    E. none of these

2. $3y + 5y = (3 + 5)y$ is an example of which property?

    A. associative

    B. commutative

    C. distributive

    D. identity for addition

    E. none of these

Homework Review Quick Check

Simplify the expression.

3. $(91a + 56) + (a + 4)$ 4. $\dfrac{2}{3}d - \dfrac{1}{6}d - \dfrac{1}{2}d$ 5. $x + x + x + x + x$

NAME _____ DATE _____

Practice

For use with Lesson 2.3: Simplifying Expressions in Two Variables

Identify the like terms in each expression.

1. $6x^2 + 5x + x$

2. $-7xy - 3y + xy + 5x$

3. $11a^3 - a^2 + 20a^3 - 9a^2$

4. $8 + 4b + 3b$

5. $cd^5 + 2cd - c^2d$

6. $(14x^2 + 3xy - y^2) + (xy + 7)$

Simplify by combining like terms.

7. $72x + y + x$

8. $-4y - xy + xy$

9. $15a^2 + 6a + 17a^2$

10. $-3c - 13d + 10c - 7d$

11. $105 + cd + 6d$

12. $7m - 2m + 8n + 5n$

13. $33b^3 + 9b^2 - 2b - 9b^2$

14. $c^2 + c^2 + c^2$

15. $2x - 5y - 5x + y$

16. $x^2y + 6xy + 3xy - 1$

17. $(-9m - 0.67n) + (m + 4n)$

18. $59x^3 + 45x^2 + 21x$

19. $\frac{7}{3}k + \frac{1}{6}l - k$

20. $172x - y + \frac{5}{9}y$

21. $m^2 - 7mn + 8m^2 + 3mn - 4n^2$

22. $0.8r^2 - 7.7r^2$

23. $-5391.26z + 17.53z + 0.165$

24. $54.17x^2 + 3x - 2.76x - 2x^2$

25. $\frac{35}{4} + xy + \frac{1}{16}xy$

26. $-\frac{11}{8}u^3v - \frac{5}{4}uv^3 - \frac{9}{8}u^3v$

27. $65\frac{1}{7} + c - \frac{8}{7} + d$

28. $(6g - h) - (-h + 6g)$

29. $B + 2A + 2A + B + A$

30. $-2t^2 + 2.11t + (-0.07t)$

31. Find a simplified algebraic expression for the perimeter of the triangle below.

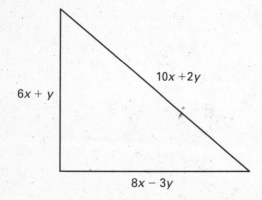

6x + y

10x +2y

8x − 3y

NAME _____ DATE _____

Using Properties with Expressions

GOAL Use properties to simplify expressions.

> The properties of addition and multiplication with the
> distributive property are used to simplify algebraic expressions
> containing multiple operations.

Understanding the Main Ideas

Knowledge of mathematical properties can make it possible to simplify algebraic
expressions without using a calculator.

EXAMPLE 1 _____

Identify the properties used in simplifying the algebraic expression.

a. $(4x + 5y - x)(8x + y - 8x - y)$ **b.** $[(5a + 7b) + 13b] - 5(a - 2b)$

SOLUTION

Step	Property
a. $(4x + 5y - x)(8x + y - 8x - y)$	
$= (4x - x + 5y)(8x - 8x + y - y)$	Commutative property of addition
$= (4x - x + 5y)(0)$	Addition property of zero
$= 0$	Multiplication property of zero
b. $[(5a + 7b) + 13b] - 5(a - 2b)$	
$= [5a + (7b + 13b)] - 5(a - 2b)$	Associative property of addition
$= 5a + 20b - 5a + 10b$	Distributive property
$= 5a - 5a + 20b + 10b$	Commutative property of addition
$= 0 + 30b$	Addition property of zero
$= 30b$	Identity property of addition

Name the mathematical property that justifies each step in the simplification.

1. $(17 + 8y - 16 - 8y)(x + x)$

     $= (17 + 8y - 16 - 8y)[(1 + 1)x]$ **a.** _____

     $= (17 - 16 + 8y - 8y)[(1 + 1)x]$ **b.** _____

     $= (1 + 0)(2x)$ **c.** _____

     $= (1)(2x)$ **d.** _____

     $= 2x$ **e.** _____

(continued)

NAME _____ DATE _____

Using Properties with Expressions

2. $13[6x^2 + (x^2 - x)] + 4x^2$

$$= 13[(6x^2 + x^2) - x] + 4x^2 \qquad \textbf{a.} \ \underline{\hspace{3cm}}$$

$$= 13[(6 + 1)x^2 - x] + 4x^2 \qquad \textbf{b.} \ \underline{\hspace{3cm}}$$

$$= 13[7x^2 - x] + 4x^2 \qquad \text{Simplify.}$$

$$= 91x^2 - 13x + 4x^2 \qquad \textbf{c.} \ \underline{\hspace{3cm}}$$

$$= 91x^2 + 4x^2 - 13x \qquad \textbf{d.} \ \underline{\hspace{3cm}}$$

$$= (91 + 4)x^2 - 13x \qquad \textbf{e.} \ \underline{\hspace{3cm}}$$

$$= 95x^2 - 13x \qquad \text{Simplify.}$$

Having quick access to the mathematical properties shortens the time necessary in simplifying algebraic expressions.

Show the steps and state the mathematical properties used in simplifying the expression.

3. $11k^2 - 15k + 8(k^2 + 2k)$ 4. $9[(a^2 + ab) + 2ab] - 3(9ab + b^2)$

5. $[-3 - (x - 3)](2x + y - y - 2x)$

Mixed Review

6. Evaluate: $|3 - 8| - 10\left|\dfrac{3}{5}\right|$

7. Simplify: $x + x^2 + x^2 - x$

8. Evaluate: $10 + 2(ab + a - ab)$ when $a = 0.09$ and $b = 165.002$

9. Evaluate: $\left(\dfrac{7}{11} - \dfrac{3}{22}\right)^3$

NAME _____ DATE _____

Quick Check

Review of Topic 2, Lesson 3

Standardized Testing Quick Check

1. What are the like terms in the expression $3x^2 + 3x + 7x^2 - 3$?

    A. $3x^2$ and $3x$

    B. $3x^2$, $3x$, and -3

    C. $3x^2$ and $7x^2$

    D. $3x$ and -3

    E. none of these

2. Which expression below *can* be simplified?

    A. $7x + 7x$

    B. $7x + 7$

    C. $7x^2 + 7x$

    D. A, B, and C

    E. none of these

Homework Review Quick Check

Simplify the expression.

3. $7 - \dfrac{1}{2}(c^2 + cd - cd + d^2) + 20$

4. $m^5n + mn^5 + 13mn^5$

5. $100(0.01x^2 - y^2) + (x^2 - 0.02y^2)$

NAME _____ DATE _____

Practice

For use with Lesson 2.4: Using Properties with Expressions

Identify the mathematical property illustrated in each identity.

1. $x + x^2 + 5x = x + 5x + x^2$

2. $-7 + 14(-3y + y) = -7 + -42y + 14y$

3. $51ab - 51ab = 0$

4. $(7.6 + 4.9b) + 3.6b = 7.6 + (4.9b + 3.6b)$

5. $(c^2d^5 + 2c)(1) = c^2d^5 + 2c$

6. $(x^2 + xy + y^2)(0)(xy) = 0$

State the mathematical property used in each step of the simplification.

7. $(y + 4xy - 11 - y)(\frac{1}{3} + x + \frac{2}{3} - x)$

$= (y - y + 4xy - 11)(\frac{1}{3} + \frac{2}{3} + x - x)$ **a.** _____

$= (0 + 4xy - 11)(1 + 0)$ **b.** _____

$= (4xy - 11)(1)$ **c.** _____

$= 4xy - 11$ **d.** _____

8. $\dfrac{[c^3 + (c^2 - c^3) - c^2]}{5(c - 1)} = \dfrac{[c^3 + (-c^3 + c^2) - c^2]}{5(c - 1)}$ **a.** _____

$= \dfrac{[c^3 + (-c^3 + c^2) - c^2]}{5c - 5}$ **b.** _____

$= \dfrac{[(c^3 - c^3) + (c^2 - c^2)]}{5c - 5}$ **c.** _____

$= \dfrac{0}{5c - 5}$ or 0, if $c \neq \dfrac{1}{5}$ **d.** _____

9. $10a + (6b - 10a) + b = (6b - 10a) + 10a + b$ **a.** _____

$= 6b + (-10a + 10a) + b$ **b.** _____

$= 6b + (0) + b$ **c.** _____

$= 6b + b$ **d.** _____

$= 7b$ Simplify.

Show the steps and state the mathematical properties used in simplifying the expression.

10. $14(x - 1) - 7(2x - 2)$

11. $\left(\dfrac{1}{2} + x + \dfrac{1}{2} - x\right)(3x + y + 2x)$

12. $[7uv + (v - 7uv)][(u + v) - (u + v)]$

13. $16 - 5(x + 4)$

14. $11c + d + c + d - 2d$

15. $(g + 6h) + (43h + 23g) + 89g$

NAME _____ DATE _____

Assessment

For use with Topic 2: Algebraic Expressions

Evaluate the expression when $x = -\frac{3}{4}$ and $y = 8$.

1. $12x$

2. $-|2x| + y$

3. $16(x - 3)$

4. $y^2 - 32x^2$

5. $-3y \div x$

6. $\frac{2}{3}x + \frac{1}{4}y$

Simplify by combining like terms.

7. $-20x + 3x$

8. $4b - 13b + 10$

9. $-6.2a - 7 + 8.1a$

10. $\frac{2}{5}y - y + \frac{3}{5}y$

11. $2\frac{1}{3} + \frac{5}{6}x + \frac{1}{2}x$

12. $7(-3y + 2) - 10$

13. $y + 4x - 3y$

14. $-3xy - 5y + 7xy$

15. $2x^2y^2 + 4x^2y + 8x^2y^2$

16. $23x - 2y + \frac{5}{8}y$

17. $1.5a^2 - 3.8a^2$

18. $-8m^2 + 3mn - n^2 - 5mn$

Identify the mathematical property illustrated in each identity.

19. $5 - 8x + 2 = 5 + 2 - 8x$

20. $-7x + 2(x - 5) = -7x + 2x - 10$

21. $(x + 2) + 9 = x + (2 + 9)$

22. $(x^2 - xy^2)(1) = x^2 - xy^2$

State the mathematical property used in each step of the simplification.

23. $12x + 3(7 - 4x) = 12x + 21 - 12x$ **a.** _____

$= 12x - 12x + 21$ **b.** _____

$= 21$ **c.** _____

Show the steps and state the mathematical properties used in simplifying the expression.

24. $[2x - (y + 2x)][(-3 + x) - x + 4]$

25. $-16(x - 2) + 8(2x - 4)$

26. Find the perimeter of a rectangle with length 22.4 meters and width 10.6 meters.

27. Use the formula for distance, $d = rt$, where r is the rate and t is time to find the rate of a car which travels 100 miles in 1.5 hours.

28. Find a simplified algebraic expression for the perimeter of the figure below.

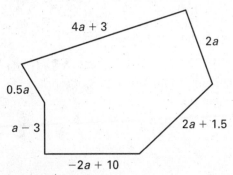

4a + 3

2a

0.5a

a − 3

2a + 1.5

−2a + 10

LESSON

3.1

Solving Equations

GOAL Solve linear equations.

A linear equation is formed when two linear expressions are connected by an equal sign.

Terms to Know   ## Example/Illustration

Linear Equation equation in which the variable is raised to the first power and does not occur in a denominator, inside a square root symbol, or inside an absolute value symbol	**Linear equation** $x + 3 = 7$ $4y - 8 = 19$ $3x + 4 + x = 4(x + 1)$ $5a = a + 24$	**Not a linear equation** $x^2 - 2 = 0$ $\lvert x + 2 \rvert = 5$ $\dfrac{2}{x} = 7$ $\sqrt{x - 4} = 16$
Solution of a linear equation a number that, when substituted for the variable in a linear equation, results in a true statement	$x = 4$ is the solution of $x + 3 = 7$ since $4 + 3 = 7 \Rightarrow 7 = 7.$	
Equivalent Equations equations with the same solution(s)	$x + 3 = 7$ and $x = 4$ $4y - 8 = 19$ and $4y = 27$ $3x + 4 + 6x = 4(x + 1)$ and $9x + 4 = 4x + 4$	

Understanding the Main Ideas

You can solve an equation by writing an equivalent equation that has the variable alone, or isolated, on one side. Any change, or transformation you apply to one side of the equation must be done to the other side to keep the equation in balance. The transformations are based on the rules of algebra called the properties of equality.

> **Properties of Equality**
>
> **Addition Property of Equality**   If $a = b$, then $a + c = b + c.$
>
> **Subtraction Property of Equality**   If $a = b$, then $a - c = b - c.$
>
> **Multiplication Property of Equality** If $a = b$, then $ca = cb.$
>
> **Division Property of Equality**   If $a = b$ and $c \neq 0$, then $\dfrac{a}{c} = \dfrac{b}{c}.$

Inverse operations are used to isolate the variable in an equation. The appropriate property of equality keeps the balance of the equation.

(continued)

NAME _____ DATE _____

Solving Equations

EXAMPLE 1

Solve each equation using the properties of equality.

a. $x + 6 = 5$ **b.** $y - 7 = 3$ **c.** $-11a = 1$ **d.** $\dfrac{b}{9} = -1$

SOLUTION

a. $x + 6 = 5$ Write original equation.

$x + 6 - 6 = 5 - 6$ Subtraction property of equality

$x = -1$ Simplify.

b. $y - 7 = 3$ Write original equation.

$y - 7 + 7 = 3 + 7$ Addition property of equality

$y = 10$ Simplify.

c. $-11a = 1$ Write original equation.

$\dfrac{-11a}{-11} = \dfrac{1}{-11}$ Division property of equality

$a = -\dfrac{1}{11}$ Simplify.

d. $\dfrac{b}{9} = -1$ Write original equation.

$9\left(\dfrac{b}{9}\right) = 9(-1)$ Multiplication property of equality

$b = -9$ Simplify.

Solve the equation.

1. $x - 21 = -20$ **2.** $6k = 3$ **3.** $B + 100 = 95$ **4.** $\dfrac{y}{-7} = \dfrac{-8}{49}$

When solving linear equations that require two or more transformations, simplify one or both sides of the equation (if needed) and use inverse operations to isolate the variable.

(continued)

Topic 3

Solving Equations

EXAMPLE 2 _____

Solve each equation using the properties of equality.

a. $34x - 3 = 31$ 　　　　　　　　　**b.** $\dfrac{y}{13} + 8 = 2$

SOLUTION

a. 　　$34x - 3 = 31$ 　　　　　　　Write original equation.

$34x - 3 + 3 = 31 + 3$ 　　　　　Addition property of equality

$34x = 34$ 　　　　　　　　Simplify.

$\dfrac{34x}{34} = \dfrac{34}{34}$ 　　　　　　Division property of equality

$x = 1$ 　　　　　　　　　Simplify.

b. 　　$\dfrac{y}{13} + 8 = 2$ 　　　　　　　Write original equation.

$\dfrac{y}{13} + 8 - 8 = 2 - 8$ 　　　　　Subtraction property of equality

$\dfrac{y}{13} = -6$ 　　　　　　　Simplify.

$13\left(\dfrac{y}{13}\right) = 13(-6)$ 　　　　　Multiplication property of equality

$y = -78$ 　　　　　　　Simplify.

Solve the equation.

5. $\dfrac{x}{-15} - 11 = -1$ 　　　　**6.** $6c + 6 = 0$ 　　　　**7.** $\dfrac{m}{0.1} - 8.7 = 6.9$

8. $\dfrac{4}{3}y + 5 = 13$ 　 *Hint:* dividing by $\dfrac{4}{3}$ gives the same result as multiplying by $\dfrac{3}{4}$.

Some linear equations have variables on both sides. Use the strategy of collecting the variables on the side with the greater variable coefficient. If either side of the equation can be simplified, do the simplifications before using inverse operations.

(continued)

NAME _____ DATE _____

Solving Equations

EXAMPLE 3

Solve the equation.

a. $16 - 6x = x + 2$

b. $7 + 9(y + 1) = y + 8 + 4y$

SOLUTION

a.

$16 - 6x = x + 2$	Write original equation.
$16 - 6x + 6x = x + 6x + 2$	Add $6x$ to each side.
$16 = 7x + 2$	Simplify.
$16 - 2 = 7x + 2 - 2$	Subtract 2 from each side.
$14 = 7x$	Simplify.
$\dfrac{14}{7} = \dfrac{7x}{7}$	Divide each side by 7.
$2 = x$	Simplify.

b.

$7 + 9(y + 1) = y + 8 + 4y$	Write original equation.
$7 + 9y + 9 = 5y + 8$	Simplify.
$9y + 16 = 5y + 8$	Simplify.
$9y - 5y + 16 = 5y - 5y + 8$	Subtract $5y$ from each side.
$4y + 16 = 8$	Simplify.
$4y + 16 - 16 = 8 - 16$	Subtract 16 from each side.
$4y = -8$	Simplify.
$\dfrac{4y}{4} = \dfrac{-8}{4}$	Divide each side by 4.
$y = -2$	Simplify.

Solve the equation.

9. $2c + 45 = c$

10. $19x - 4 = 3x + 4$

11. $32y + 8y = -120$

12. $5(x + 2) - (x - 12) = -3x + 5$

Mixed Review

13. Simplify: $-1 - 1 - 1(x - y)$

14. Evaluate $|2a + b|$ when $a = \frac{1}{2}$ and $b = -3$.

15. Find the perimeter of a square with side length $7x - 3$.

Topic 3

Quick Check

Review of Topic 2, Lesson 4

Standardized Testing Quick Check

1. Which expression is equivalent to $5 - 3(x + 2)$?

    A. $2x + 4$

    B. $11 - 3x$

    C. $-3x - 1$

    D. $7 - 3x$

    E. none of these

2. Which expression is equivalent to $(9 + 2)(-6xy + 6xy)$?

    A. 1

    B. 0

    C. $-54xy + 12xy$

    D. -42

    E. none of these

Homework Review Quick Check

State the mathematical property illustrated in each identity.

3. $(-67) + (1 - 43) = (1 - 43) + (-67)$

4. $3 + (-3) = 0$

5. $[12 + (-35)] + (-2) = 12 + [(-35) + (-2)]$

Geometry
Basic Skills Workbook: Diagnosis and Remediation

NAME _____ DATE _____

Practice

For use with Lesson 3.1: Solving Equations

Fill in the blank(s) with the property of equality used in the simplification.

1. $x + 5 = 10$ Original equation

$x + 5 - 5 = 10 - 5$ _____

$x = 5$ Simplify.

2. $3d = 6$ Original equation

$\dfrac{3d}{6} = \dfrac{6}{3}$ _____

$d = 2$ Simplify.

3. $\dfrac{a}{-5} = \dfrac{1}{2}$ Original equation

$-5\left(\dfrac{a}{-5}\right) = -5\left(\dfrac{1}{2}\right)$ _____

$a = -\dfrac{5}{2}$ Simplify.

4. $3y - 1 = 4$ Original equation

$3y - 1 + 1 = 4 + 1$ _____

$3y = 5$ Simplify.

$\dfrac{3y}{3} = \dfrac{5}{3}$ _____

$y = \dfrac{5}{3}$ Simplify.

Solve the equation.

5. $\dfrac{d}{8} = -4$

6. $d + 8 = -4$

7. $d - 8 = -4$

8. $8d = -4$

9. $17x - 1 = 16$

10. $-1 = 7y + 6$

11. $\dfrac{k}{89} = 0$

12. $6.05p + p = 14.1$

13. $43x = x$

14. $\dfrac{12}{7}h - \dfrac{6}{7} = \dfrac{1}{3}$

15. $2(x - 4) = x - (3x + 3)$

16. $8.17 = a + 0.03 + a$

17. $1 + 7(3t + 2) = t - 5$

18. $-r + 78 = 34$

19. $\dfrac{x}{-7} + 8 + 3x = 10$

20. *Dog pen* The perimeter of a rectangular dog pen is 24 feet. The length is twice the width. What are the dimensions of the dog pen?

Geometry
Basic Skills Workbook: Diagnosis and Remediation

47

Topic 3

NAME _____ DATE _____

Solving Inequalities

GOAL **Solve linear inequalities.**

> Solving linear inequalities in one variable is much like solving
> linear equations in one variable.

Terms to Know	Example/Illustration
Linear inequality an open sentence formed when an inequality symbol is placed between two linear expressions	$x + 3 < 7$ $\quad$ $4y - 8 \geq 9$ $5a \leq a + 24$ $3x + 4 + 6x > 4(x + 1)$
Solution of a linear inequality a number that, when substituted for the variable in the inequality, results in a true statement	$x = 3$ is a solution of $x + 3 < 7$ since $3 + 3 < 7 \Longrightarrow 6 < 7$.

Understanding the Main Ideas

Solving a linear inequality involves rewriting the inequalities into equivalent
inequalities until the variable is isolated. The transformations used are similar to
those used with equations except for two important differences. When you multiply or divide by a negative number, you must reverse the inequality symbol to
maintain a true statement. For example, 2 is less than 4, but $(-2)(2)$ is not less
than $(-2)(4)$ because -4 is greater than -8. The table below summarizes the
transformations that can be used.

Transformations that Produce Equivalent Inequalities	
1. **Add the same number to *each* side.**	If $a < b$, then $a + c < b + c$ If $a > b$, then $a + c > b + c$
2. **Subtract the same number from *each* side.**	If $a < b$, then $a - c < b - c$ If $a > b$, then $a - c > b - c$
3. **Multiply each side by the same *positive* number.**	If $a < b$ and $c > 0$ then $ca < cb$ If $a > b$ and $c > 0$ then $ca > cb$
4. **Divide each side by the same *positive* number.**	If $a < b$ and $c > 0$ then $\dfrac{a}{c} < \dfrac{b}{c}$ If $a > b$ and $c > 0$ then $\dfrac{a}{c} > \dfrac{b}{c}$
5. **Multiply each side by the same *negative* number and reverse the inequality symbol.**	If $a < b$ and $c < 0$ then $ca > cb$ If $a > b$ and $c < 0$ then $ca < cb$
6. **Divide both sides by the same *negative* number and reverse the inequality symbol.**	If $a < b$ and $c < 0$ then $\dfrac{a}{c} > \dfrac{b}{c}$ If $a > b$ and $c < 0$ then $\dfrac{a}{c} < \dfrac{b}{c}$

(continued)

NAME _____ DATE _____

Solving Inequalities

Be careful to consider each step in your solution. Checking a few values in the original inequality is a good check of your answer.

EXAMPLE 1 _____

Solve each inequality.

a. $y - 63 \geq 81$ **b.** $-a + 5 < 1$ **c.** $\dfrac{b}{14} - 2 \leq 0$

SOLUTION

a.
$$y - 63 \geq 81$$ Write original inequality.
$$y - 63 + 63 \geq 81 + 63$$ Add 63 to each side.
$$y \geq 144$$ Simplify.

b.
$$-a + 5 < 1$$ Write original inequality.
$$-a + 5 - 5 < 1 - 5$$ Subtract 5 from each side.
$$-a < -4$$ Simplify.
$$\dfrac{-a}{-1} > \dfrac{-4}{-1}$$ Divide each side by -1 and *reverse* the inequality
$$a > 4$$ Simplify.

c.
$$\dfrac{b}{14} - 2 \leq 0$$ Write original inequality.
$$\dfrac{b}{14} - 2 + 2 \leq 0 + 2$$ Add 2 to each side.
$$\dfrac{b}{14} \leq 2$$ Simplify.
$$14 \cdot \dfrac{b}{14} \leq 14 \cdot 2$$ Multiply each side by 14.
$$b \leq 28$$ Simplify.

Solve the inequality.

1. $10x + 1 > -2$ **2.** $d - 13 \leq 99$ **3.** $7.86 - 4.8y \geq 0.5$ **4.** $\dfrac{x}{-3} < 11$

In solving multi-step linear inequalities, mathematical properties are used to simplify each side of the inequality before applying the transformations to isolate the variable.

(continued)

Topic 3

NAME _____ DATE _____

Solving Inequalities

EXAMPLE 2

Solve each inequality.

a. $7(x + 15) - (2x - 9) > x + 4 + 3x$ **b.** $\frac{1}{3}y + \left(-\frac{4}{3}y + 6\right) \le 27$

SOLUTION

a. $7(x + 15) - (2x - 9) > x + 4 + 3x$ Write original inequality.

$\qquad 7x + 105 - 2x + 9 > x + 4 + 3x$ Use distributive property.

$\qquad 7x - 2x + 105 + 9 > x + 3x + 4$ Use commutative property.

$\qquad\qquad\quad 5x + 114 > 4x + 4$ Simplify.

$\qquad 5x - 4x + 114 > 4x - 4x + 4$ Subtract $4x$ from each side.

$\qquad\qquad\qquad x + 114 > 4$ Simplify.

$\qquad x + 114 - 114 > 4 - 114$ Subtract 114 from each side.

$\qquad\qquad\qquad\qquad x > -110$ Simplify.

b. $\frac{1}{3}y + \left(-\frac{4}{3}y + 6\right) \le 27$ Write original inequality.

$\qquad \left(\frac{1}{3}y - \frac{4}{3}y\right) + 6 \le 27$ Use associative property.

$\qquad\qquad (-y) + 6 \le 27$ Simplify.

$\qquad (-y) + 6 - 6 \le 27 - 6$ Subtract 6 from each side.

$\qquad\qquad\qquad -y \le 21$ Simplify.

$\qquad\qquad\quad \frac{-y}{-1} \ge \frac{21}{-1}$ Divide each side by -1 and *reverse* the inequality.

$\qquad\qquad\qquad y \ge -21$ Simplify.

Solve the inequality.

5. $19x - 5(7 - 3x) < -1$ **6.** $41.2y + (3.8y + 3) \ge -42$

7. $-99 - a + 99 > 4a$ **8.** $b + b + b \ge (1253)(0)(342,678)$

(continued)

Topic 3

Solving Inequalities

Mixed Review

9. Evaluate: $\dfrac{41}{5} + \dfrac{1}{10} - 3\left(\dfrac{4}{5}\right)$

10. Simplify: $3x^6 + x^5 + 4x^5$

11. Solve: $5y - y = 3$

12. Evaluate $-5a^4 + 1000b$ when $a = -1$ and $b = 0.007$.

13. Subtract: $\dfrac{1}{10} - \left|\dfrac{11}{10}\right|$

14. Solve: $\dfrac{5c}{8} = 1$

15. Simplify: $7(4d - 5) - (d - 90)$

Quick Check

Review of Topic 3, Lesson 1

Standardized Testing Quick Check

1. What is the solution of the equation $12x - 24 = 12$?

   A. $x = 25$

   B. $x = -1$

   C. $x = 3$

   D. $x = -3$

   E. none of these

2. What is the first step used to solve $3x + 1 = 4$?

   A. Add 1 to each side.

   B. Subtract 1 from each side.

   C. Multiply each side by 3.

   D. Divide each side by 3.

   E. none of these

Homework Review Quick Check

Solve the equation.

3. $-81y = (3)^3$

4. $16x + 22 = 54$

5. $c + c + c = 2c - 7$

NAME _____ DATE _____

Practice

For use with Lesson 3.2: Solving Inequalities

Check to see if the number given is a solution of the inequality.

1. $-x > 15, -19$

2. $6y + 4 \geq 0, -0.1$

3. $4c \leq 3c, \dfrac{1}{2}$

4. $9 - (x + 3) < 1, 0$

Solve the inequality.

5. $\dfrac{x}{3} \leq -1$

6. $d + 3 > -1$

7. $d - 3 < -1$

8. $3d > -1$

9. $-x - 1 \geq 1$

10. $-11 > 13y + 6$

11. $\dfrac{k}{-7} + \dfrac{5}{6} \geq \dfrac{-1}{6}$

12. $5.05b + 0.95b > 12$

13. $63x < x$

14. $\dfrac{1}{6}h - 8 \geq \dfrac{1}{6}$

15. $-(x - 4) + x < 3x + 3$

16. $n + 0.27 > 9n + 0.03$

17. $31 + 2(t + 2) < -5$

18. $-8x \geq 3$

19. $\dfrac{6x}{-14} + 7 > 10$

20. $-77k + |-9 - 6| < 0$

21. $2(-5)^2 + y > 100$

22. $0.1(100x + 1000) > 1000$

23. $8.25a - 8a + 13.12 > -0.13$

24. $\dfrac{1}{2}(m - 6) + \dfrac{3}{2}(m - 16) \leq m - 9$

25. *Perimeter*   The perimeter of the figure below is at least 10 meters.
  Find the possible values of x.

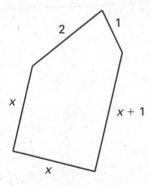

Geometry   53
Basic Skills Workbook: Diagnosis and Remediation

Topic 3

NAME _____   DATE _____

Proportions

GOAL **Solve proportions.**

> Many real world problems can be solved using proportions. For
> example, the amount of medicine a doctor prescribes to a patient is
> in proportion to the patient's weight.

Terms to Know	**Example/Illustration**
Ratio of *a* to *b* the relationship $\frac{a}{b}$ of two quantities *a* and *b* that are measured in the same units (Since the units are the same, they are left off and the ratio is reduced.)	Ratios can be written as $a : b$, a to b, and $\frac{a}{b}$. 3 ft : 2 ft $\Longrightarrow$ 3 : 2 4 boys to 3 girls $\Longrightarrow$ 4 to 3 $\frac{10 \text{ years}}{40 \text{ years}} \Longrightarrow \frac{10}{40}$ or $\frac{1}{4}$
Proportion equation that states that two ratios are equal. For example, $\frac{a}{b} = \frac{c}{d}$, where a, b, c and d are nonzero real numbers	$\dfrac{\text{known conversion weight}}{\text{patient's weight}}$ $= \dfrac{\text{known dosage}}{\text{patient's dosage}}$ $\dfrac{140}{110} = \dfrac{0.625}{x}$

Understanding the Main Ideas

The equating of two ratios forms a proportion. For example $\frac{a}{b} = \frac{c}{d}$ is a

proportion if the original fractions were ratios. It is read "*a* is to *b* as *c* is to *d*." The numbers *a* and *d* are called the *extremes* of the proportion. The numbers *b* and *c* are called the *means* of the proportion. The cross product property will be used when solving for a variable in a proportion.

> **Cross Product Property**
>
> The product of the extremes equals the product of the means.
>
> If $\frac{a}{b} = \frac{c}{d}$, then $ad = bc$. **Example:** $\frac{2}{3} = \frac{4}{6} \Longrightarrow 2 \cdot 6 = 3 \cdot 4$

Be careful that you only use the cross property when you start with a proportion.

EXAMPLE 1 _____

Solve each proportion.

a. $\dfrac{x}{55} = \dfrac{-1}{5}$ **b.** $\dfrac{9}{35} = \dfrac{4}{y}$

(continued)

NAME _____ DATE _____

Proportions

SOLUTION

a. $\dfrac{x}{55} = \dfrac{-1}{5}$ Write original proportion.

     $5x = -55$ Use cross product property.

     $x = -11$ Divide each side by 5.

b. $\dfrac{9}{35} = \dfrac{4}{y}$ Write original proportion.

     $9y = 140$ Use cross product property.

     $y = \dfrac{140}{9}$ Divide each side by 9.

Solve the proportion.

1. $\dfrac{x}{6} = \dfrac{1}{24}$ **2.** $\dfrac{4}{d} = \dfrac{32}{72}$ **3.** $\dfrac{1}{17} = \dfrac{y}{-34}$ **4.** $\dfrac{-5}{3} = \dfrac{8}{a}$

You may need to use the mathematical properties to solve more complicated proportions.

EXAMPLE 2 _____

Solve the proportion.

a. $\dfrac{x}{x + 2} = \dfrac{7}{6}$ b. $\dfrac{-9}{14} = \dfrac{3y - 1}{2}$

SOLUTION

a. $\dfrac{x}{x + 2} = \dfrac{7}{6}$ Write the original proportion.

     $7(x + 2) = 6x$ Use cross product property.

     $7x + 14 = 6x$ Use distributive property.

   $7x - 6x + 14 = 6x - 6x$ Subtract $6x$ from each side.

     $x + 14 = 0$ Simplify.

   $x + 14 - 14 = 0 - 14$ Subtract 14 from each side.

     $x = -14$ Simplify.

(continued)

Topic 3

Geometry   **55**
Basic Skills Workbook: Diagnosis and Remediation

NAME _____ DATE _____

Proportions

b. $\dfrac{-9}{14} = \dfrac{3y - 1}{2}$ Write original proportion.

$14(3y - 1) = -18$ Use cross product property.

$42y - 14 = -18$ Use distributive property.

$42y - 14 + 14 = -18 + 14$ Add 14 to each side.

$42y = -4$ Simplify.

$\dfrac{42y}{42} = \dfrac{-4}{42}$ Divide each side by 42.

$y = -\dfrac{2}{21}$ Simplify.

Solve the proportion.

5. $\dfrac{8}{4} = \dfrac{x - 3}{3}$ **6.** $\dfrac{-10}{7} = \dfrac{2}{y + 1}$ **7.** $\dfrac{10}{a} = \dfrac{4}{a + 0.1}$ **8.** $\dfrac{x}{5x - 11} = \dfrac{3}{4}$

Mixed Review

Solve.

9. $\dfrac{c}{5} + 17 = 2$

10. $-v > 1.3$

11. $4 - (c - 8) = 13$

Geometry
Basic Skills Workbook: Diagnosis and Remediation

NAME _____ DATE _____

Quick Check

Review of Topic 3, Lesson 2

Standardized Testing Quick Check

1. For which inequality is the integer -8 a solution?

   A. $x + 1 > 0$

   B. $x - 1 > 0$

   C. $-x > 0$

   D. $x > 0$

   E. none of these

2. An inequality is reversed if you

   A. add -1 to each side.

   B. subtract 1 from each side.

   C. multiply each side by $\frac{1}{2}$.

   D. divide each side by -1.

   E. none of these

Homework Review Quick Check

Solve the inequality.

3. $79 - 4x + 1$

4. $10 - 6(x - 5) \geq 0$

5. $b > 2b - 3$

Topic 3

Practice

For use with Lesson 3.3: Proportions

1. Write the extremes and the means of the proportion $\frac{2}{6} = \frac{1}{3}$.

Write yes or no to tell whether the equation is equivalent to $\dfrac{a}{b} = \dfrac{c}{d}$.

2. $ad = bc$

3. $ab = cd$

4. $\dfrac{b}{a} = \dfrac{d}{c}$

Solve the proportion.

5. $\dfrac{x}{8} = \dfrac{1}{2}$

6. $\dfrac{8}{x} = \dfrac{1}{2}$

7. $\dfrac{8}{11} = \dfrac{x}{2}$

8. $\dfrac{8}{11} = \dfrac{1}{x}$

9. $\dfrac{y}{-5} = \dfrac{10}{-2}$

10. $\dfrac{k}{-45} = \dfrac{7}{-5}$

11. $\dfrac{6}{b} = \dfrac{15}{3}$

12. $\dfrac{9}{100} = \dfrac{7.2}{x}$

13. $\dfrac{3}{75} = \dfrac{p}{100}$

14. $\dfrac{y}{-18} = \dfrac{-11}{-6}$

15. $\dfrac{36}{k} = \dfrac{6}{-7}$

16. $\dfrac{125}{100} = \dfrac{x}{4}$

17. $\dfrac{x}{-3} = \dfrac{x+1}{-2}$

18. $\dfrac{h}{87} = \dfrac{h-1}{7}$

19. $\dfrac{8y}{-5} = \dfrac{10y+3}{1}$

20. $\dfrac{70}{s+68} = \dfrac{1}{s-1}$

21. $\dfrac{r+7}{r} = \dfrac{5}{12}$

22. $\dfrac{3}{2} = \dfrac{w+3}{7w+2}$

23. *Scale drawing*   In a scale drawing one inch represents one and one-half meters. How many meters does 8 inches represent?

24. *Recipe*   Mary is baking bread. The recipe says that 12 cups of flour will make 3 loaves of bread. How many cups are needed for 2 loaves of bread?

25. *Medicine dosage*   If the correct medicine dosage for a 140-pound woman is 0.625 milligrams, what is the correct dosage for a 155-pound woman?

Topic 3

Solving Problems

GOAL Solve problems using equations, inequalities, and proportions.

> Equations, inequalities, and proportions can model real-life situations.

Verbal Model ↓	Ask yourself what you need to know to solve the problem. Then write a verbal model that will give you what you need to know.
Labels ↓	Assign labels to each part of your verbal model.
Algebraic Model ↓	Use the labels to write an algebraic model based on your verbal model.
Solve ↓	Solve the algebraic model and answer the original question.
Check	Check that your answer is reasonable.

Understanding the Main Ideas

Writing algebraic expressions, equations, or inequalities that represent real-life situations is called modeling. The expression, equation, or inequality is a mathematical model of the real-life situation. When writing and solving a mathematical model, use the problem solving plan shown above.

EXAMPLE 1 _____

Your final average in algebra class is 80% of your semester average plus 20% of your final exam grade. If your semester average is 89 points and your final average is 90 points, what is your final exam grade?

SOLUTION

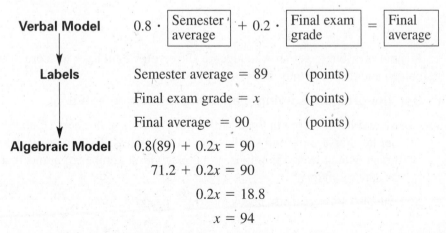

Verbal Model $0.8 \cdot \boxed{\text{Semester average}} + 0.2 \cdot \boxed{\text{Final exam grade}} = \boxed{\text{Final average}}$

Labels
Semester average $= 89$ (points)
Final exam grade $= x$ (points)
Final average $= 90$ (points)

Algebraic Model
$$0.8(89) + 0.2x = 90$$
$$71.2 + 0.2x = 90$$
$$0.2x = 18.8$$
$$x = 94$$

Your final exam grade is 94 points.

(continued)

Topic 3

Solving Problems

Use the problem solving plan using models to solve.

1. The high school marching band has a bake sale to raise funds for a competition. Cookies are 25¢ and brownies are 50¢. If they sell the same amount of each and make $56.25, how many brownies are sold?

2. You join a CD club. The beginning membership is $25. CDs are all the same price, which includes tax. What is the price of a CD if you buy 5 when you send in your membership and write a check for $95.65?

Some algebraic models become inequalities.

EXAMPLE 2

Your semester average in algebra class is 89 before you take the final exam. The semester average counts for 70% of your final grade and the final exam counts for 30%. Can you receive an A in algebra if the lowest A is 93 and the highest A is 100?

SOLUTION

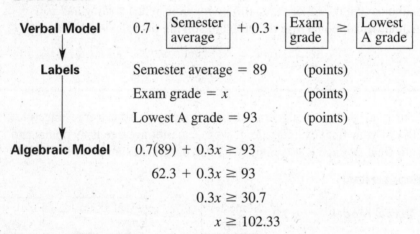

Verbal Model $0.7 \cdot \boxed{\begin{array}{c}\text{Semester} \\ \text{average}\end{array}} + 0.3 \cdot \boxed{\begin{array}{c}\text{Exam} \\ \text{grade}\end{array}} \geq \boxed{\begin{array}{c}\text{Lowest} \\ \text{A grade}\end{array}}$

Labels   Semester average = 89   (points)

Exam grade = x (points)

Lowest A grade = 93   (points)

Algebraic Model $0.7(89) + 0.3x \geq 93$

$62.3 + 0.3x \geq 93$

$0.3x \geq 30.7$

$x \geq 102.33$

It is not possible to get an A in algebra since you would have to score over 102 points, and there are only 100 points possible.

Use the problem solving plan using models to solve.

3. Meredith hopes to win the award for high points in her horse club. She is going to her last horse show and she has earned 134 points. To be sure of victory she must finish with at least 155 points. What is the least number of points that she can earn to be sure of victory?

You can also solve real-life problems involving proportions.

(continued)

Topic 3

NAME _____ DATE _____

Solving Problems

EXAMPLE 3

If you use an average of 4 pencils in one week, on average, how many pencils will you use in one year?

SOLUTION

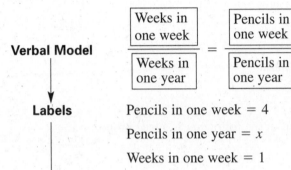

Verbal Model

$$\frac{\text{Weeks in one week}}{\text{Weeks in one year}} = \frac{\text{Pencils in one week}}{\text{Pencils in one year}}$$

Labels

Pencils in one week $= 4$ (pencils)

Pencils in one year $= x$ (pencils)

Weeks in one week $= 1$ (weeks)

Weeks in one year $= 52$ (weeks)

Algebraic Model $\dfrac{1}{52} = \dfrac{4}{x}$

$x = 4 \cdot 52$ Use cross product property.

$x = 208$

You use 208 pencils in one year.

Use the problem solving plan and models to solve.

4. If 2 pizzas are needed for 5 people, what is the least number of pizzas that must be ordered for 23 people?

5. Your dog eats $1\frac{2}{3}$ cups of dog food per day. How much dog food is needed to feed your dog for a month?

Mixed Review

Solve.

6. $51x - 11 = 40$ **7.** $-y - y + 9 = -65$ **8.** $4 - 10c > 13$

Simplify the expression.

9. $734.13 - (-23.7)$ **10.** $3(a + 9b) + 17(2a + 2b)$

11. $\dfrac{5}{9} \cdot \dfrac{72}{25}$ **12.** $|-4.9| + |8.7|$

13. Evaluate $-x - (3)^2$ when $x = -19$.

Topic 3

Geometry   **61**
Basic Skills Workbook: Diagnosis and Remediation

NAME _____ DATE _____

Quick Check

Review of Topic 3, Lesson 3

Standardized Testing Quick Check

1. Which of the following is a ratio?

   A. $\dfrac{25 \text{ miles}}{1 \text{ hour}}$

   B. 2 inches to 1 mile

   C. $\dfrac{a}{b} = \dfrac{c}{d}$

   D. 10 feet : 20 feet

   E. none of these

2. If $ae = gh$, which of the following is a true proportion?

   A. $\dfrac{a}{e} = \dfrac{g}{h}$

   B. $\dfrac{a}{g} = \dfrac{e}{h}$

   C. $\dfrac{a}{h} = \dfrac{g}{e}$

   D. $\dfrac{e}{a} = \dfrac{g}{h}$

   E. none of these

Homework Review Quick Check

Solve the proportion.

3. $\dfrac{x}{20} = \dfrac{3}{30}$

4. $\dfrac{4}{y} = \dfrac{1}{17}$

5. $\dfrac{7}{3} = \dfrac{p}{100}$

6. $\dfrac{25}{5} = \dfrac{9}{x + 2}$

7. $\dfrac{y + 4}{3} = \dfrac{y}{5}$

8. $\dfrac{6}{3} = \dfrac{x + 8}{-1}$

NAME _____ DATE _____

Practice

For use with Lesson 3.4: Solving Problems

Use the problem solving plan and models to solve.

1. ***T-shirts***   Your baseball team buys 10 T-shirts that are all the same price. Your bill is $159.00, including 6% tax. What is the price of one T-shirt before tax is added?

2. ***Stuffed animals***   Alice has collected two stuffed animals each week for 20 weeks. How many more weeks must she continue to collect at this rate to have more than 100 stuffed animals?

3. ***Paint***   One gallon of paint will cover up to 400 square feet of previously painted surfaces. You buy paint to repaint 6 rectangular walls that are each 15 feet long and 10 feet high. How many gallons of paint do you need to buy? (*Hint*: you cannot buy part of a gallon.)

4. ***Fabric***   Mrs. Hayes is cutting fabric to make kites. Each kite takes $\frac{1}{2}$ yard of fabric. She cuts 30 yards into $\frac{1}{2}$ yard lengths. How many kites can she make?

5. ***Flowers***   You can plant 42 flowers in 1 square foot of garden space. How many flowers should you buy for a circular space with radius 8 inches?

6. ***Basketball***   You scored 24 points in your last basketball game. You made 5 free throws at 1 point each and one 3-point field goal. The rest of your points were from 2-point field goals. How many 2-point field goals did you make?

7. ***Tiles***   Jim is installing ceramic tile behind his kitchen sink. Each tile is a square with side length $4\frac{1}{4}$ inches. For each tile, he allows an additional $\frac{1}{8}$ inch per side for the grout between tiles. How many tiles should he buy if the space measures 2.1 square feet?

8. ***Final grade***   Your final math grade is 80% of the semester average and 20% of your final exam grade. Your grades during the semester are 72, 78, 78, and 86. You need a C (77 to 84) or better to play sports in the fall. What is the lowest score you can receive on your exam to be able to play sports?

9. ***Fish***   When placing fish into an aquarium, there should be no more than one inch of fish length per one gallon of water. Carl has 12 inches of fish length in a 30-gallon tank. When he goes to buy more fish, at most, how many 2.5-inch fish can he buy?

10. ***Dance***   Your class has hired a disc jockey for a Friday night dance. The disc jockey costs $250. Three hundred people have signed up to attend the dance. How much should you charge for tickets in order to make a profit of more than $500?

Topic 3

NAME _____ DATE _____

Assessment

For use with Topic 3: Equations and Inequalities

Solve the equation.

1. $21x - 12 = 30$

2. $-3 = 5 - 2d$

3. $\dfrac{m}{8} = -11$

4. $-13 = 52y$

5. $\dfrac{5}{4}x - 3 = 2$

6. $0.21p = 2.42 - p$

7. $2z + 0.98 = z - 0.52$

8. $3(6t - 2) - 8 = 5 - t$

9. $12 - 3x = -51$

Check to see whether the number given is a solution of the inequality.

10. $5s - 7 \le 12, 4$

11. $-3b \ge 25, -9$

12. $12 - (4 - x) > 0, -8$

Solve the inequality.

13. $n - 6 < -2$

14. $2.5k + 3 \ge 4.5$

15. $\dfrac{c}{4} + \dfrac{1}{9} > -\dfrac{8}{9}$

16. $-20 \ge 11y + 13$

17. $17 - 5d \le 2$

18. $\dfrac{4h}{-5} - 4 \ge 16$

19. $16 - 2(r - 5) < -6$

20. $4(-2)^2 + b > 36$

21. $-6.25c + 4c - 4.5 < 0$

Solve the proportion.

22. $\dfrac{x}{10} = \dfrac{3}{2}$

23. $\dfrac{15}{y} = \dfrac{3}{4}$

24. $\dfrac{16}{p} = \dfrac{64}{4}$

25. $\dfrac{c}{-72} = \dfrac{2}{-3}$

26. $\dfrac{4}{-50} = \dfrac{z}{-125}$

27. $\dfrac{54}{t} = \dfrac{3.75}{2.5}$

28. $\dfrac{-1.8}{m} = \dfrac{-1.5}{6}$

29. $\dfrac{3 + x}{x} = \dfrac{9}{8}$

30. $\dfrac{5}{2} = \dfrac{t + 2}{t - 1}$

Use the problem-solving plan to solve.

31. *Word Game*   In a word game, you spell a word using 7 letters for a score of 14 points. You use 3 vowels worth 1 point each and an N worth 2 points. The other 3 letters are M's. How many points is an M worth?

32. *Candles*   You are making candles to sell at the school craft fair. It costs $2.25 to make each candle, and you sell the candles for $7.75. You want to make a profit of $99. How many candles must you sell?

33. *Jeans*   You buy 2 pairs of jeans at the same time. The total amount you pay is $54.06, including 6% sales tax. What is the price of one pair of jeans before sales tax is added?

NAME _____ DATE _____

Drawing and Measuring Angles

GOAL **Draw, measure, and identify angles.**

> Accurate measurement of angles can help in solving real-life problems such as finding the height of a mountain.

Terms to Know / Example/Illustration

Terms to Know	Example/Illustration
Ray part of a line that consists of a point, called an *initial point*, and all points on the line that extend in one direction	ray *AB*, or $\overrightarrow{AB}$
Angle consists of two different rays that have the same initial point. The rays are the *sides* of the angle, and the initial point is the *vertex* of the angle.	This angle has sides $\overrightarrow{AB}$ and $\overrightarrow{AC}$, and it is denoted by $\angle BAC$, $\angle CAB$, or $\angle A$.
Acute Angle angle with measure between 0° and 90°	$\angle A$ measures less than 90° and is an acute angle.
Right Angle angle with measure equal to 90°	$\angle C$ measures 90° and is a right angle.
Obtuse Angle angle with measure between 90° and 180°	$\angle B$ measures more than 90° and is an obtuse angle.
Straight Angle angle with measure equal to 180°	$\angle CBA$ measures 180° and is a straight angle.

Understanding the Main Ideas

The ancient Babylonians used a base 60 number system. They divided a full circle into 360 degrees. One degree is $\frac{1}{360}$ of the way around the circle. We continue to use degree measurement today.

The measurement of $\angle A$ is denoted by $m\angle A$. Angles can be measured by using a protractor by placing the center of the base at the vertex of the angle with one ray along the base. Extend the other ray until you can read the measurement. Most protractors read both right to left and left to right. Read from the ray along the base to the other ray starting with 0°.

(continued)

Topic 4

NAME _____ DATE _____

Drawing and Measuring Angles

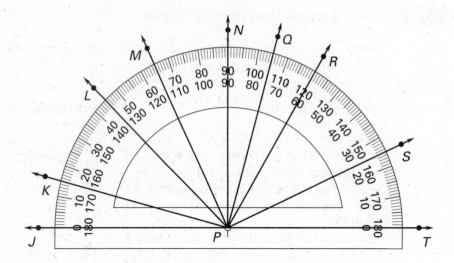

EXAMPLE 1

Identify 8 angles above and use the protractor to read their degree meas-
urements. Also give the type of angle (acute, right, obtuse, or straight).

SAMPLE SOLUTION

	Angle		Measure	Type
	Angle		**Measure**	**Type**
a.	∠TPS	Measure from ray $\overrightarrow{PT}$ to ray $\overrightarrow{PS}$ right to left; 0° starts on **bottom.**	25°	acute
b.	∠TPR	Measure from ray $\overrightarrow{PT}$ to ray $\overrightarrow{PR}$ right to left; 0° starts on **bottom.**	60°	acute
c.	∠TPN	Measure from ray $\overrightarrow{PT}$ to ray $\overrightarrow{PN}$ right to left; 0° starts on **bottom.**	90°	right
d.	∠TPL	Measure from ray $\overrightarrow{PT}$ to ray $\overrightarrow{PL}$ right to left; 0° starts on **bottom.**	135°	obtuse
e.	∠JPL	Measure from ray $\overrightarrow{PJ}$ to ray $\overrightarrow{PL}$ left to right; 0° starts on **top.**	45°	acute
f.	∠JPN	Measure from ray $\overrightarrow{PJ}$ to ray $\overrightarrow{PN}$ left to right; 0° starts on **top.**	90°	right
g.	∠JPR	Measure from ray $\overrightarrow{PJ}$ to ray $\overrightarrow{PR}$ left to right; 0° starts on **top.**	120°	obtuse
h.	∠JPT	Measure from ray $\overrightarrow{PJ}$ to ray $\overrightarrow{PT}$ left to right; 0° starts on **top.**	180°	straight
		or		
		Measure from ray $\overrightarrow{PT}$ to ray $\overrightarrow{PJ}$ right to left; 0° starts on **bottom.**	180°	straight

(continued)

Geometry
Basic Skills Workbook: Diagnosis and Remediation

NAME _____ DATE _____

Drawing and Measuring Angles

In Exercises 1–4, use the protractor and angles in Example 1 to find the measure and type of each angle.

1. $\angle TPQ$ **2.** $\angle JPS$ **3.** $\angle JPK$ **4.** $\angle TPJ$

A protractor can also be used to draw an angle. Draw a ray from the center point (vertex) along the edge to the right or left. Then mark the degree measurement desired using the bottom numbers if the first ray points to the right and the top numbers if the first ray points to the left. Draw a ray from the vertex to the mark.

Use the protractor to draw the angles with the given measure.

5. $55°$ **6.** $115°$ **7.** $5°$ **8.** $107°$

Mixed Review

Simplify the expression.

9. $\dfrac{4}{7} \div 4 \cdot (-7)$ **10.** $8uv + 19v - 4(2uv + v)$ **11.** $|-1 - 9| \cdot |3 - 7|$

Solve the equation or inequality.

12. $\dfrac{7}{100} = \dfrac{5}{x}$ **13.** $6s = s + 3$ **14.** $3x + 5 \geq 9x - 7$

15. Evaluate $-(xy)^2 + y$ when $x = -8$ and $y = \dfrac{1}{4}$.

NAME _____ DATE _____

Quick Check

Review of Topic 3, Lesson 4

Standardized Testing Quick Check

1. For every teacher at your high school, there are 12 students. If *t* represents the number of teachers, which of the following could represent the number of students?

   A. $12t$ **B.** $t + 12$

   C. $t - 12$ **D.** $\dfrac{t}{12}$

   E. none of these

2. Jane's dinner cost $8.50 including a 15% tip. Which equation could be solved to find the amount of the bill *b* before the tip?

   A. $0.15b = \$8.50$ **B.** $\$8.50 + 0.15b = b$

   C. $b + 0.15b = \$8.50$ **D.** $(b)(0.15b) = \$8.50$

   E. none of these

Homework Review Quick Check

Use the problem-solving plan and models to solve.

3. *College Students*   At the local college there are 2 male engineering students for each female engineering student. If there were 600 male students in the freshman class last year, how many students were there in all?

4. *Class Election*   The junior class is electing class officers. You are running for president. You know that you have support of one fourth of the 610-member class. If there is one other candidate running for president, what is the least number of additional votes you must get to be assured of a win?

5. *Clarinet Practice*   Brian spends 10 hours per week practicing the clarinet. He wants to raise his time spent practicing per week by 20%. How many hours a week must he add?

NAME _____ DATE _____

Practice

For use with Lesson 4.1: Drawing and Measuring Angles

**Use a protractor to find the measure of ∠A to the nearest degree.
Then identify the angle as acute, right, obtuse, or straight.**

1.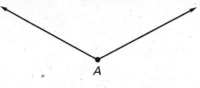

2.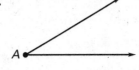

3.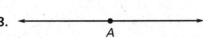

4.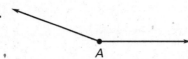

5.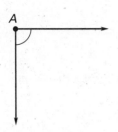

6.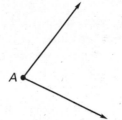

**Use a protractor to draw an angle with the degree measurement
given and label the angle appropriately.**

7. $m\angle G = 70°$ **8.** $m\angle CDE = 130°$ **9.** $m\angle K = 45°$ **10.** $m\angle M = 177°$

11. $m\angle B = 15°$ **12.** $m\angle W = 98°$ **13.** $m\angle JKL = 154°$ **14.** $m\angle A = 86°$

Topic 4

NAME _____ DATE _____

Polygons

GOAL   **Define and identify basic polygons.**

> Polygons such as triangles, rectangles, and squares occur often in the natural world.

Terms to Know	Example/Illustration
Parallel Lines two lines that are coplanar and do not intersect	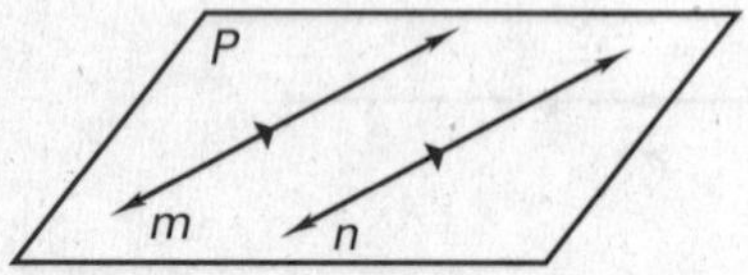 $m \parallel n$ The arrows in the diagram indicate parallel lines.
Perpendicular Lines two lines that intersect to form a right angle	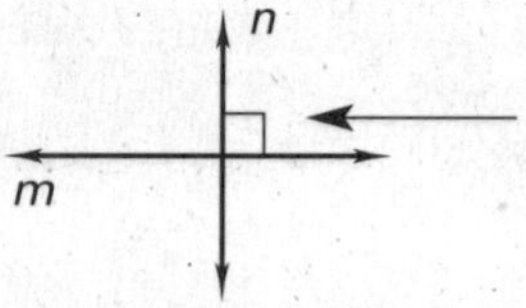 Indicates a right angle.
Line Segment part of a line consisting of two endpoints and all the points between them	 Line segment AB is symbolized by $\overline{AB}$.
Polygon a closed plane figure whose sides are straight line segments that intersect at their endpoints	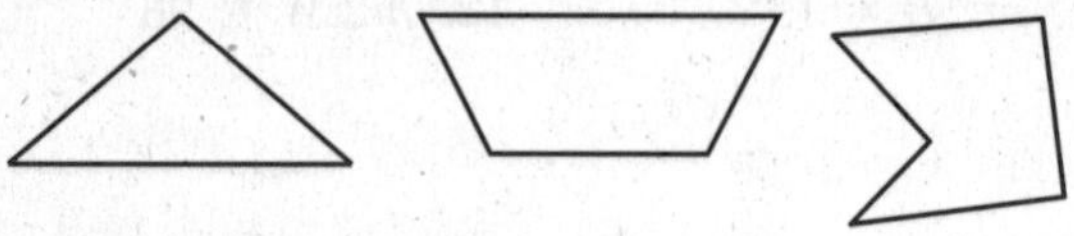
Regular Polygon polygon with equal side lengths and equal angle measures	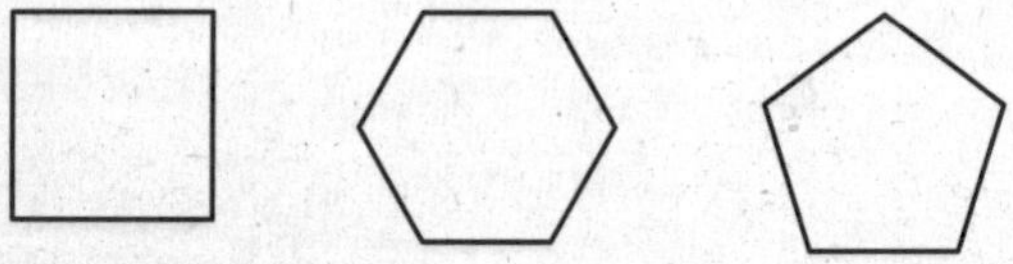
Triangle polygon with three sides	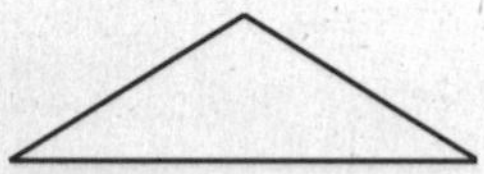
Quadrilateral polygon with four sides	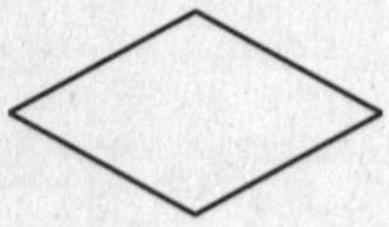

Understanding the Main Ideas

A triangle can be classified by its sides and angles as shown in the diagrams on the next page. In the diagrams, matching *congruent marks* identify angles and segments that are equal in measure, or congruent.

(continued)

Topic 4

NAME _____ DATE _____

Polygons

NAMES OF TRIANGLES

Classification by Sides

Equilateral Triangle	**Isosceles Triangle**	**Scalene Triangle**
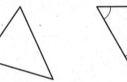	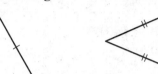	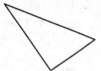
3 congruent sides	At least 2 congruent sides	No congruent sides

Classification by Angles

Acute Triangle	**Equiangular Triangle**	**Right Triangle**	**Obtuse Triangle**
3 acute angles	3 congruent angles	1 right angle	1 obtuse angle

Note: An equiangular triangle is also acute.

EXAMPLE 1

Sketch the indicated type of triangle.

a. right scalene   **b.** isosceles obtuse

SOLUTION

a. right scalene

b. isosceles obtuse

Sketch the indicated type of triangle.

1. right isosceles   **2.** equilateral

3. scalene acute   **4.** obtuse scalene

(continued)

Topic 4

NAME _____ DATE _____

Polygons

There are many ways to classify quadrilaterals. Five are shown below.

NAMING QUADRILATERALS

1. **Parallelogram**
   quadrilateral with opposite sides parallel

2. **Rectangle**
   parallelogram with 4 right angles

   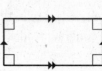

3. **Square**
   rectangle with 4 congruent sides
   Note: The square is the regular
   quadrilateral.

4. **Rhombus**
   parallelogram with 4 congruent sides

5. **Trapezoid**
   quadrilateral with only one pair
   of parallel sides

Give the name that best describes each quadrilateral.

5.   6.   7. 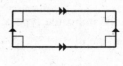   8.

Mixed Review

Simplify the expression.

9. $(|-2^3|)^4$

10. $7 - \dfrac{1}{3}$

11. $100.98 \div 0.002$

12. $7.1xy + 0.036 + 1.25 + x$

Solve the equation.

13. $45y + 2y = y + 23$

14. $\dfrac{t}{t+3} = \dfrac{-1}{8}$

15. Evaluate $17 + 13(x - 1)$ when $x = 1$.

NAME _____ DATE _____

Quick Check

Review of Topic 4, Lesson 1

Standardized Testing Quick Check

1. What are the sides of $\angle XYZ$?

   A. $\overrightarrow{XY}$ and $\overrightarrow{YZ}$ **B.** $\overrightarrow{YX}$ and $\overrightarrow{YZ}$

   C. $\overrightarrow{XY}$ and $\overrightarrow{XZ}$ **D.** $\overrightarrow{ZX}$ and $\overrightarrow{ZY}$

   E. none of these

2. The measure of an obtuse angle is

   A. 90°.   **B.** between 0° and 90°.

   C. 180°.   **D.** between 90° and 180°.

   E. none of these

Homework Review Quick Check

Use a protractor to find $m\angle A$. Identify $\angle A$ as acute, right, obtuse, or straight.

3.

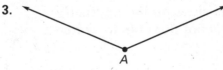

4.

5.

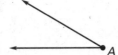

Geometry
Basic Skills Workbook: Diagnosis and Remediation

Topic 4

NAME _____ DATE _____

Practice

For use with Lesson 4.2: Polygons

Classify the triangle by its angles and sides.

1.

2.

3.

Decide whether the statement is *true* or *false*.

4. A rectangle is sometimes a trapezoid.

5. A parallelogram is always a rhombus.

6. A square is always a rectangle.

7. A rhombus is always a square.

8. A trapezoid is always a quadrilateral.

Use the description to sketch a figure. Indicate congruent sides, congruent angles, right angles, and parallel sides on the figure. If the figure has a shorter name, give it. If it is not possible to sketch a figure, write *not possible*.

9. a regular quadrilateral

10. an equilateral right triangle

11. a parallelogram with 4 congruent angles

12. a quadrilateral with exactly one pair of parallel sides

13. a regular trapezoid

14. an isosceles triangle with one acute angle

15. a quadrilateral with all sides of different length

Topic 4

NAME _____ DATE _____

Circles

GOAL **Draw, define, and use circles.**

> To understand circles, you first need to know some special terms that
> describe a circle.

Terms to Know	Example/Illustration
Circle set of all points in a plane that are the same distance from a given point called the center	center
Diameter distance across the circle, through its center	diameter
Radius distance from the center of the circle to a point on the circle (The radius is one half the diameter.)	radius

Understanding the Main Ideas

To draw a circle, you must first locate where you want the center of the circle to
lie. The points of the circle will all be an equal distance from this point. You may
use circular objects such as jar lids to draw circles, but if you want to construct a
circle of a particular size, a compass can be used.

To draw a circle with a compass, open the compass and set the distance between
the sharp point and the point of the pencil for the length of the radius desired.
You can measure to assure the correct length. Place the sharp point of the compass on the location of the center of your circle and rotate around this point using
the pencil to draw your circle (see figure below).

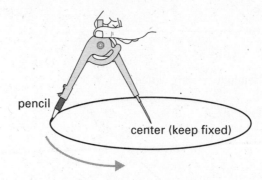

pencil

center (keep fixed)

(continued)

Topic 4

NAME _____ DATE _____

Circles

EXAMPLE 1

Use a compass to draw a circle with diameter 4 inches.

SOLUTION

Since a compass measures from the center to a point on the circle, you must know the length of the radius. The radius is one half the length of the diameter. So, the radius is $\frac{1}{2}$(4 inches) = 2 inches. Open the compass and set the distance between the sharp point and the pencil point for 2 inches. Measure to be sure of accuracy. Rotate the pencil around the center holding the center point steady. Be careful not to slip off the center.

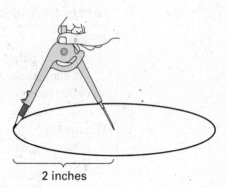

2 inches

Use a compass to draw a circle with the given diameter or radius.

1. radius = 5 cm

2. diameter = 1 in.

3. diameter = 4.6 cm

4. radius = $1\frac{1}{4}$ in.

Mixed Review

5. Draw a straight angle and label it.

6. Simplify: $12 \cdot \frac{1}{5} + \frac{3}{5} - \frac{3}{10} \cdot \left(\frac{1}{10}\right)^2$

7. Solve: $8.65 - 100x \geq 13.6$

8. Write an expression for the perimeter of the triangle below.

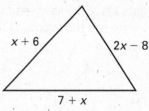

$x + 6$ $2x - 8$

$7 + x$

9. Evaluate $\dfrac{x - 17}{x + 17}$ when $x = 5$.

10. Find $m\angle K$.

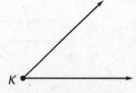

K

Topic 4

NAME _____ DATE _____

Quick Check

Review of Topic 4, Lesson 2

Standardized Testing Quick Check

1. Which of the following cannot complete the sentence to make a true statement? A square is a _____.

   A. rectangle   **B.** rhombus

   C. quadrilateral   **D.** parallelogram

   E. none of these

2. An equilateral triangle is also all of the following *except*

   A. an acute triangle.   **B.** a right triangle.

   C. an isosceles triangle.   **D.** a regular polygon.

   E. none of these

Homework Review Quick Check

Give the name of the polygon with the given description.

3. a quadrilateral with exactly one pair of parallel sides

4. a triangle with at least 2 congruent sides

5. a parallelogram with all four angles equal in measure

Topic 4

NAME _____   DATE _____

Practice

For use with Lesson 4.3: Circles

Find the diameter of a circle with the given radius.

1. 10 mm

2. 17 in.

3. 2 ft

4. $\frac{6}{7}$ m

5. 5.3 cm

6. $9\frac{1}{16}$ in.

Find the radius of a circle with the given diameter.

7. 18 ft

8. 6 in.

9. 11 miles

10. $\frac{3}{4}$ in.

11. 0.1 m

12. $44\frac{3}{8}$ in.

Use a compass to draw a circle with the given radius or diameter.

13. radius = 4 cm

14. radius = 1 in.

15. radius = $\frac{5}{8}$ in.

16. radius = $2\frac{1}{2}$ in.

17. diameter = 35 mm

18. diameter = 8.9 cm

19. diameter = $3\frac{3}{4}$ in.

20. radius = 7.5 cm

Use a compass to draw a circle and label it to match the description. Use the scale of 1 inch = 1 foot in your sketch.

21. Delores is planting a circular garden with diameter 5 feet.

22. The portholes of Bill's boat have a radius of $\frac{3}{4}$ foot.

NAME _____ DATE _____

Solids

GOAL **Define and recognize parts of solids.**

> You can use polyhedrons to solve real-life problems, such as finding the number of edges of a soccer ball.

Terms to Know	*Example/Illustration*
Polyhedron a solid that is bounded by polygons, called *faces*, that enclose a single region of space	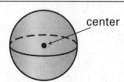
Edge a line segment formed by the intersection of two faces of a polyhedron	
Vertex point where three or more edges of a polyhedron meet	
Prism polyhedron with two congruent faces, called *bases*, that lie in parallel planes	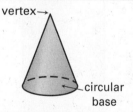
Cylinder solid with congruent circular bases that lie in parallel planes	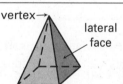
Pyramid polyhedron in which the base is a polygon and the *lateral faces* are triangles with a common *vertex*	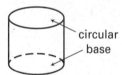
Cone solid with a circular *base* and a *vertex* that is not in the same plane as the base	
Sphere locus of points in space that are a given distance from a point, called the *center* of the sphere	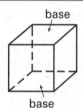

(continued)

Topic 4

NAME _____ DATE _____

Solids

Understanding the Main Ideas

It is important to be able to recognize the faces, vertices, and edges of a polyhedron.

EXAMPLE 1

Use the diagram of the polyhedron at the right.

a. List the faces.

b. List the vertices.

c. List the edges.

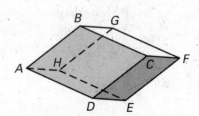

SOLUTION

a. The faces are *ABCD, CDEF, EFGH, GHAB, BCFG,* and *ADEH.*

b. The vertices are *A, B, C, D, E, F, G,* and *H.*

c. The edges are $\overline{AB}, \overline{BC}, \overline{CD}, \overline{DA}, \overline{BG}, \overline{GH}, \overline{HA}, \overline{GF}, \overline{FE}, \overline{EH}, \overline{CF},$ and $\overline{DE}.$

In Exercises 1–3, use the diagram of the polyhedron at the right.

1. List the faces.

2. List the vertices.

3. List the edges.

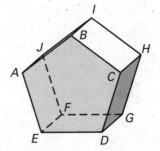

EXAMPLE 2

Decide whether the solid is a polyhedron. If so, count the number of faces, vertices, and edges of the polyhedron.

a. 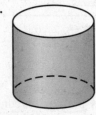 **b.** 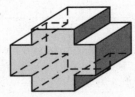 **c.**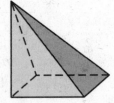

SOLUTION

a. This is not a polyhedron because it is not bounded by polygons.

b. This is a polyhedron. It has 14 faces, 24 vertices, and 36 edges.

c. This is a polyhedron. It has 5 faces, 5 vertices, and 8 edges.

(continued)

Geometry
Basic Skills Workbook: Diagnosis and Remediation

Topic 4

NAME _____ DATE _____

Solids

Decide whether the solid is a polyhedron. If so, count the number of faces, vertices, and edges of the polyhedron.

4.

5.

6.
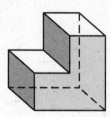

EXAMPLE 3

Identify the solid. Is it a polyhedron?

a.

b.

c.

SOLUTION

a. This is a sphere. It is *not* a polyhedron because it is not bounded by polygons.

b. This is a cone because it has a circular base and a vertex that is not in the same plane as the base. It is *not* a polyhedron.

c. This is a cylinder because it has congruent circular bases that lie in parallel planes. It is *not* a polyhedron.

Identify the solid. Is it a polyhedron?

7.
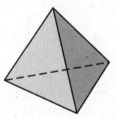

8.
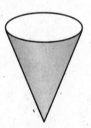

9.
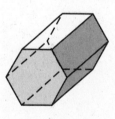

Mixed Review

Fill in each blank to make a true statement.

10. An isosceles triangle has at least _____ congruent sides.

11. An obtuse angle measures between _____ ° and _____ °.

Decide whether the statement is *sometimes*, *always*, or *never* true.

12. The absolute value of a negative number is negative.

13. A rectangle has 4 congruent sides.

Topic 4

LESSON
4.4

NAME ______________________ DATE __________

Quick Check

Review of Topic 4, Lesson 3

Standardized Testing Quick Check

1. The distance across a circle, through its center is called

A. the radius. **B.** the major arc.

C. the diameter. **D.** the tangent.

E. none of these

2. If the diameter of a circle is 6.82 centimeters, what is the radius?

A. 3.41 cm **B.** 13.64 cm

C. 46.5 cm^2 **D.** 2.61 cm

E. none of these

Homework Review Quick Check

Use a compass to draw a circle with the given radius or diameter.

3. radius = 3.6 cm **4.** diameter = 3 in.

5. radius = 20 mm

Topic 4

NAME _____ DATE _____

Practice

For use with Lesson 4.4: Solids

In Exercises 1–4, use the prism at the right.

1. List the edges.

2. List the vertices.

3. List the faces.

4. List the bases.

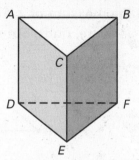

Count the number of faces, vertices, and edges of the polyhedron.

5.

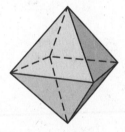

6.

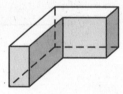

7.

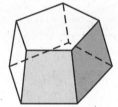

8.

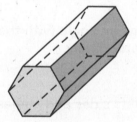

Identify the solid.

9.

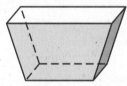

10.

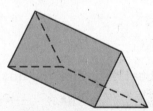

11.

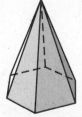

12.

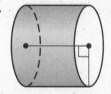

Topic 4

NAME _____ DATE _____

Assessment

For use with Topic 4: Geometric Figures and Solids

Use a protractor to find the measure of ∠A to the nearest degree. Then identify the angle as acute, right, obtuse, or straight.

1.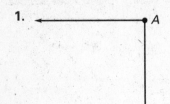

2.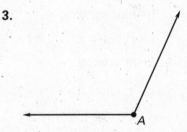

3.

4.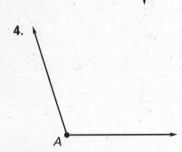

5.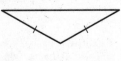

6.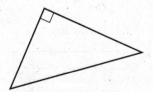

Use a protractor to draw an angle with the degree measure given and label the angle appropriately.

7. $m\angle B = 68°$

8. $m\angle JKL = 165°$

9. $m\angle X = 52°$

Classify the triangle by its sides and angles.

10.

11.

12.

Give the name that best describes the quadrilateral.

13.

14.

15.

(continued)

NAME _____ DATE _____

Assessment

For use with Topic 4: Geometric Figures and Solids

Use a compass to draw a circle with the given radius or diameter. Then complete the statement.

16. radius = 4 cm

diameter = ____?____

17. diameter = 2 in.

radius = ____?____

18. radius = $3\frac{1}{2}$ in.

diameter = ____?____

Identify the solid.

19.

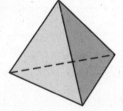

20.

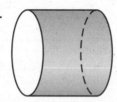

21.

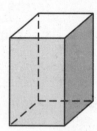

NAME _____ DATE _____

Plotting Points

GOAL | **Graph and identify points in the coordinate plane.**

> Ordered pairs of real numbers are used to represent real-life relation-
> ships. Graphing the ordered pairs on a coordinate plane gives a visual
> picture of this relationship.

Terms to Know	***Example/Illustration***
Coordinate Plane plane formed by two real number lines intersecting at a right angle	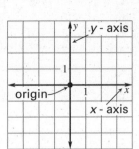
Origin point in a coordinate plane at which the horizontal axis intersects the vertical axis. The point (0, 0).	
***x*-axis** horizontal axis in a coordinate plane	
***y*-axis** vertical axis in a coordinate plane	
Ordered Pair pair of real numbers used to locate each point in the coordinate plane. The first number in an ordered pair is called the ***x*-coordinate** and the second number is called the ***y*-coordinate**	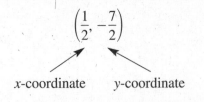

Understanding the Main Ideas

The coordinate plane extends one dimensional graphing into two dimensions. When
you are plotting points in a coordinate plane, all plotting starts at the origin (0, 0).

EXAMPLE 1 _____

Plot each point on a coordinate plane.

   a. $A(3, 1)$ **b.** $B(-5, -2)$ **c.** $C(-4, 5)$ **d.** $D(0, 3)$

(continued)

NAME _____ DATE _____

Plotting Points

SOLUTION

a. Point *A* has an *x*-coordinate of 3 and a *y*-coordinate of 1. Starting at the origin (0, 0), move 3 spaces to the right (a positive move) and 1 space up (a positive move). Place a dot on the graph to represent the point and label it *A*.

b. Point *B* has an *x*-coordinate of −5 and a *y*-coordinate of −2. Starting at the origin (0, 0), move 5 spaces to the left (a negative move) and 2 spaces down (a negative move). Place a dot on the graph to represent the point and label it *B*.

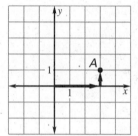
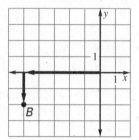

c. Point *C* has an *x*-coordinate of −4 and a *y*-coordinate of 5. Starting at the origin (0, 0), move 4 spaces to the left (a negative move) and 5 spaces up (a positive move). Place a dot on the graph to represent the point and label it *C*.

d. Point *D* has an *x*-coordinate of 0 and a *y*-coordinate of 3. Starting at the origin (0, 0), move 0 spaces to the right or left and 3 spaces up (a positive move). Place a dot on the graph to represent the point and label it *D*.

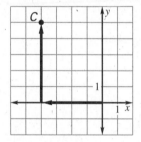
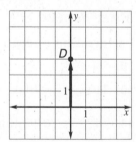

Plot each point.

1. $A(2, -1)$ **2.** $B(-3, 6)$ **3.** $C(1, 0)$ **4.** $D(5, 3)$

The axes divide the coordinate plane into four parts called **quadrants.**
All points (x, y) lie in a quadrant except for points with a 0 (zero) coordinate, which fall on either the *x*- or *y*-axis. The quadrants are labeled as shown below.

	y	
Quadrant II		Quadrant I
	1	
		1 *x*
Quadrant III		Quadrant IV

(continued)

LESSON
5.1
CONTINUED

Plotting Points

EXAMPLE 2

Give the coordinates of each point in the graph at
the right. Then give the quadrant in which the point lies.

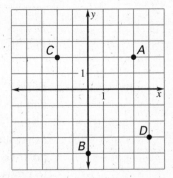

SOLUTION

		Coordinates	Quadrant
a.	A	$(4, 4)$	I
b.	B	$(-4, 4)$	II
c.	C	$(-3, 0)$	no quadrant; on x-axis
d.	D	$(2, -2)$	IV

**Give the coordinates of each point on the
graph at the right and the quadrant in
which it lies.**

5. A **6.** B

7. C **8.** D

Mixed Review
....

Perform the operation without a calculator.

9. $56(-98)$

10. $3400 + (-60) + 600 + (-140)$

11. $[(1235)(-6798)(0)(147)] \div \left(\dfrac{11}{12}\right)$

12. $\dfrac{41}{51} - \left(-\dfrac{88}{3}\right) - \dfrac{41}{51} - \dfrac{88}{3}$

Geometry
Basic Skills Workbook: Diagnosis and Remediation

NAME _____ DATE _____

Quick Check

Review of Topic 4, Lesson 4

Standardized Testing Quick Check

1. What is a solid with 2 parallel bases of unequal size called?

   A. cube   **B.** pyramid

   C. cone   **D.** prism

   E. none of these

2. The base of a pyramid must be a

   A. circle.   **B.** square.

   C. prism.   **D.** polygon.

   E. none of these

Homework Review Quick Check

Give the geometric name of the familiar solid.

3. a basketball   4. a soup can

5. a shoe box filled with pebbles and taped shut

NAME _____ DATE _____

Practice

For use with Lesson 5.1: Plotting Points

Plot and label each of the following points on the coordinate plane.

1. $A(6, -7)$

2. $B(-4, 3)$

3. $C(-5, -1)$

4. $D(0, -1)$

5. $E(2, 9)$

6. $F(-5, 0)$

7. $G(0, 0)$

8. $H(-8, -10)$

9. $I(9, 6)$

10. $J(2, -2)$

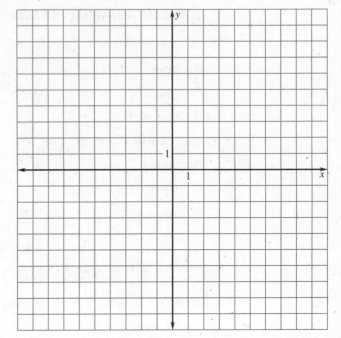

Give the coordinates of each point on the graph and the quadrant in which it lies.

11. A

12. B

13. C

14. D

15. E

16. F

17. G

18. H

19. I

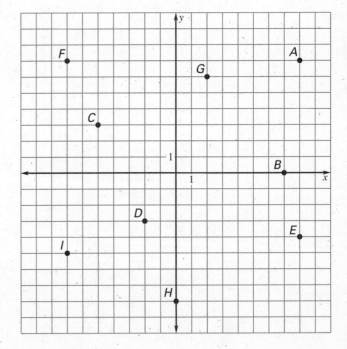

Geometry
Basic Skills Workbook: Diagnosis and Remediation

Graphing Equations

GOAL   **Graph linear equations.**

The equation $y + 3 = 5x$ is a linear equation in *two variables*. You can use a table of values to graph a linear equation in two variables.

Terms to Know	Example/Illustration
Solution of a Linear Equation an ordered pair (x, y) that makes the equation true when the values of x and y are substituted into the equation	$(1, 1)$ is a solution of $y = x$ since $1 = 1$.
Graph of an Equation in Two Variables set of *all* points (x, y) that are solutions of the equation	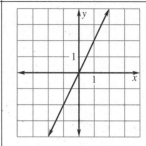

Understanding the Main Ideas

In this lesson, you will see that the graph of a *linear* equation is a line.

EXAMPLE 1

Use the graph to decide whether the point lies on the graph of $x + 2y = 4$. Justify your answer algebraically.

a. $(1, 1)$ **b.** $(-2, 3)$

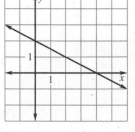

SOLUTION

a. The point $(1, 1)$ is *not* on the graph of $x + 2y = 4$. This means that $(1, 1)$ is not a solution. You can check this algebraically.

$x + 2y = 4$ Write original equation.

$1 + 2(1) \stackrel{?}{=} 4$ Substitute 1 for x and 1 for y.

$3 \neq 4$ Simplify. Not a true statement.

$(1, 1)$ is not a solution of the equation $x + 2y = 4$, so it is not on the graph.

(continued)

NAME _____ DATE _____

Graphing Equations

b. The point $(-2, 3)$ is on the graph of $x + 2y = 4$. This means that $(-2, 3)$ is a solution. You can check this algebraically.

$$x + 2y = 4 \qquad \text{Write original equation.}$$

$$-2 + 2(3) \overset{?}{=} 4 \qquad \text{Substitute } -2 \text{ for } x \text{ and } 3 \text{ for } y.$$

$$4 = 4 \qquad \text{Simplify. True statement}$$

$(-2, 3)$ is a solution of the equation $x + 2y = 4$, so it is on the graph.

Decide whether the given ordered pair is a solution of the equation.

1. $2y - 4x = 8, (-2, 8)$ **2.** $-5x - 8y = 15, (-3, 0)$

3. $y = 3x - 5, (-1, -8)$ **4.** $y - 7 = -4x, (2, 1)$

In Example 1 the point $(-2, 3)$ is on the graph of $x + 2y = 4$, but how many points does the graph have in all? The answer is that most graphs have too many points to list.

Then how can you ever graph an equation? One way is to make a table or list of a few values, plot enough solutions to recognize a pattern, and then connect the points.

GRAPHING A LINEAR EQUATION

STEP ❶ Rewrite the equation in function form, if necessary.

STEP ❷ Choose a few values of x and make a table of values.

STEP ❸ Plot the points from the table of values. A line through these points is the graph of the equation.

EXAMPLE 2

Use a table of values to graph the equation $y + 3 = 4x$.

SOLUTION

Rewrite the equation in function form by solving for y.

$$y + 3 = 4x \qquad \text{Write original equation.}$$

$$y = 4x - 3 \qquad \text{Subtract 3 from each side.}$$

Choose a few values for x and make a table of values.

Choose x.	*Substitute to find the corresponding y-value.*
-2	$y = 4(-2) - 3 = -11$
-1	$y = 4(-1) - 3 = -7$
0	$y = 4(0) - 3 = -3$
1	$y = 4(1) - 3 = 1$
2	$y = 4(2) - 3 = 5$

(continued)

Topic 5

Graphing Equations

SOLUTION-*CONTINUED*

With this table of values you have found five solutions.

$(-2, -11), (-1, -7), (0, -3), (1, 1), (2, 5)$

Plot the points. You can see that they all lie on a line.

Draw this line.

The line through the points is the graph of the equation.

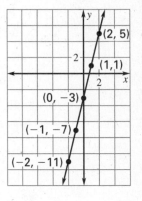

Use a table of values to graph the linear equation.

5. $y = x + 5$ **6.** $y = -4x$ **7.** $x + y = 7$ **8.** $-5x - y = 3$

Mixed Review

Solve each problem.

9. It takes 3 hours to travel 200 miles. If the same average speed is maintained, how long will it take to travel 260 miles?

10. If three times a number is $-\dfrac{14}{9}$, what is the number?

11. What is the smallest integer that satisfies the inequality $7 + 9x \geq 11$?
(*Hint:* Integers are numbers in the set . . ., $-3, -2, -1, 0, 1, 2, 3, \ldots$)

Quick Check

Review of Topic 5, Lesson 1

Standardized Testing Quick Check

1. The point $(4, -8)$ lies in which quadrant?

   A. I   **B.** II

   C. III   **D.** IV

   E. none of these

2. Each point on the coordinate plane corresponds to

   A. a real number.   **B.** an integer.

   C. an ordered pair of real numbers.

   D. four real numbers.

   E. none of these

Homework Review Quick Check

Plot the point in a coordinate plane.

3. $(5, -5)$

4. $(0, 4)$

5. $\left(-\dfrac{3}{2}, -1\right)$

NAME _____ DATE _____

Practice

For use with Lesson 5.2: Graphing Equations

Decide whether the given ordered pair is a solution of the equation.

1. $6y - 3x = -9, (2, -1)$

2. $-2x - 9y = 7, (-1, -1)$

3. $-3x + y = 12, (0, 4)$

4. $x + 4y = 48, (8, 10)$

Find three ordered pairs that are solutions of the equation.

5. $y = -2x + 12$

6. $y + 4x = -7$

Fill in the missing values in the table.

7.

x	-2	-1	0	1	2
$y = 6x + 2$	$y = 6(-2) + 2$	$y = 6(\_\_\_) + 2$	$y = $ _____	$y = $ _____	$y = $ _____
y	-10	-4	_____	_____	_____

8.

x	-14	-7	_____	_____	14
$y = \frac{2}{7}x - 4$	$y = $ _____	$y = \frac{2}{7}(\_\_\_) - 4$	$y = $ _____	$y = \frac{2}{7}(7) - 4$	$y = $ _____
y	-8	_____	-4	_____	_____

Graph the linear equation.

9. $y = x + 3$

10. $y = -3x + 2$

11. $6y = 3x + 6$

12. $x = y - 1$

13. $-2x - 2y - 2 = 2$

14. $7x + 2y = 4$

15. $x + y = 10$

16. $x - y = 0$

17. $4y = 3x + 6$

18. $9x + 9y = 18$

19. $y = 5$ *(Hint: $y = 0x + 5$; x may take on any value.)*

Figures in a Plane

GOAL **Graph polygons and their transformations in a plane.**

> Transformations occur in many real-life applications using geometric figures. Computer graphics, architecture, wall paper, and fabric designs are just a few places where geometric patterns occur.

Understanding the Main Ideas

You can graph polygons in a coordinate plane by connecting points that represent the vertices of the polygon.

EXAMPLE 1

Graph each set of polygons in a coordinate plane by connecting points that represent the vertices of the polygon.

a. $A(1, 6), B(3, 2), C(5, 4)$ **b.** $A(-4, 2), B(4, 2), C(4, -1), D(-4, -1)$

SOLUTION

a. The points form a triangle as shown.

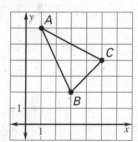

b. The points form a rectangle as shown.

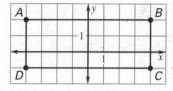

Graph each set of points, connect in order with the last back to the first, then name the polygon.

1. $A(0, 0), B(6, 0), C(0, 6)$

2. $A(-3, 3), B(3, 3), C(3, -3), D(-3, -3)$

3. $A(-5, -2), B(2, -2), C(0, 4), D(-3, 4)$

4. $A(0, 8), B(3, 6), C(3, 0), D(-3, 0), E(-3, 6)$

Geometric patterns are formed when another polygon of the same size and shape is plotted by sliding or flipping the original polygon.

(continued)

Figures in a Plane

EXAMPLE 2

Move each vertex of the triangle with the vertices $A(0, 2)$, $B(2, 6)$, and $C(4, 4)$ using the following motion rules. Describe how each new triangle is related to the original triangle.

a. Add 3 to each x-coordinate.

b. Subtract 2 from each y-coordinate.

c. Replace each x with $-x$.

SOLUTION

The original triangle with vertices $A(0, 2)$, $B(2, 6)$, and $C(4, 4)$ is shown below.

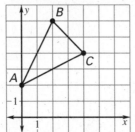

a. This triangle has the vertices $A(0 + 3, 2)$, $B(2 + 3, 6)$, and $C(4 + 3, 4)$. In other words, this triangle is shifted 3 spaces to the right of the original triangle. It has the same shape and size.

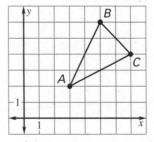

b. This triangle has the vertices $A(0, 2 - 2)$, $B(2, 6 - 2)$, and $C(4, 4 - 2)$. This triangle is shifted 2 spaces down from the original triangle. It has the same size and shape.

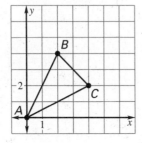

c. This triangle has the vertices $A(0, 2)$, $B(-2, 6)$, and $C(-4, 4)$. This triangle is the original triangle flipped over the y-axis. It has the same size and shape.

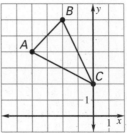

(continued)

Figures in a Plane

Graph the original polygon, move vertices using the motion rule, and then describe the new polygon in relation to the original.

5. $A(1, 4)$, $B(5, 4)$, $C(4, 1)$, $D(0, 1)$; subtract 4 from each x-coordinate.

6. $A(-3, -5)$, $B(-5, -1)$, $C(-1, 0)$; add 6 to each y-coordinate.

7. $A(7, 3)$, $B(5, 1)$, $C(-4, 1)$, $D(-2, 3)$; replace each y with $-y$.

8. $A(2, -2)$, $B(8, 1)$, $C(0, 0)$; add 1 to each x-coordinate and subtract 1 from each y-coordinate.

Mixed Review

Identify the figure described.

9. a parallelogram with 4 congruent sides

10. an angle between 0° and 90°

11. a triangle with all congruent sides

12. a rectangle with all sides congruent

13. a triangle with all angles congruent

14. a rhombus with congruent angles

15. an angle of 180°

NAME _____ DATE _____

Quick Check

Review of Topic 5, Lesson 2

Standardized Testing Quick Check

1. Which of the following is a solution of $9x + y = 2$?

    A. $(1, 11)$

    B. $\left(0, \dfrac{2}{9}\right)$

    C. $\left(\dfrac{2}{9}, 0\right)$

    D. $(9, 2)$

    E. none of these

2. $(1, 3)$ is a solution of which equation?

    A. $x + 3y = 0$

    B. $y = x + 3$

    C. $2x + 3y = 11$

    D. $x - y = 2$

    E. none of these

Homework Review Quick Check

Graph the linear equation.

3. $y = 7x - 2$

4. $x + y = 5$

5. $10x + 2y = 4$

NAME _____ DATE _____

Practice

For use with Lesson 5.3: Figures in a Plane

Graph each set of points and connect them in order with segments to form a polygon. Then name the polygon.

1. $A(0, 0)$, $B(-4, 2)$, $C(3, 5)$

2. $A(-1, 6)$, $B(1, 6)$, $C(1, -1)$, $D(-1, -1)$

3. $A(9, -7)$, $B(6, -7)$, $C(6, -4)$, $D(9, -4)$

4. $A(0, 0)$, $B(1, 5)$, $C(9, 5)$, $D(8, 0)$

5. $A(-3, -2)$, $B(1, -2)$, $C(0, 4)$, $D(-2, 4)$

6. $A(-4, 0)$, $B(0, 3)$, $C(4, 0)$, $D(0, -3)$

7. $A(10, 3)$, $B(4, 6)$, $C(0, 3)$, $D(2, 2)$

8. $A(-5, 1)$, $B(-4, 6)$, $C(1, 6)$, $D(2, 1)$

Move vertices of the polygon in the indicated exercise above using the motion rule and then describe the new polygon in relation to the original.

9. Add 4 to each x-coordinate in Exercise 1.

10. Subtract 7 from each y-coordinate in Exercise 2.

11. Replace each x with $-x$ in Exercise 3.

12. Replace each y with $-y$ in Exercise 4.

13. Add -6 to each x-coordinate in Exercise 5.

14. Add 2 to each y-coordinate in Exercise 6.

15. Replace each x with $-x$ and y with $-y$ in Exercise 7.

16. Add 1 to each x-coordinate and 3 to each y-coordinate in Exercise 8.

17. Draw a rhombus in the coordinate plane and label its vertices.

18. Draw an octagon (polygon with 8 sides) in the coordinate plane and label its vertices.

19. Draw an isosceles triangle in the coordinate plane and label its vertices.

20. Draw a polygon of your choice in the coordinate plane and label its vertices.

NAME _____ DATE _____

Perimeter and Area

GOAL Find the perimeter and area of a figure in the coordinate plane.

Terms to Know	Example/Illustration
Perimeter the sum of the lengths of the sides of a figure	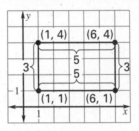 The perimeter of the rectangle is $5 + 3 + 5 + 3 = 16$ units.
Area the number of square units contained in the interior of a figure	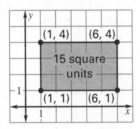 The area of the rectangle = 15 square units.

Understanding the Main Ideas

To find the **perimeter** of a figure in the coordinate plane, find the length of each side of the figure, and then add these lengths. If a figure has sides that are either vertical or horizontal, you can find the length of a side by finding the horizontal or vertical distance between the points at the vertices of the figure.

If the points of two vertices of a figure are (x_1, y) and (x_2, y), then the line is horizontal, and the horizontal distance is

$$|x_2 - x_1|.$$

Similarly, if the points of two vertices of a figure are (x, y_1) and (x, y_2), then the line is vertical, and the vertical distance is

$$|y_2 - y_1|.$$

In a coordinate plane with a grid, each square of the grid is a **unit square,** which has an area of one square unit. To find the **area** of a figure in the coordinate plane, count the number of unit squares that are in the interior of the figure.

(continued)

Perimeter and Area

EXAMPLE 1

The points $A(2, 1)$, $B(2, 4)$, $C(-3, 4)$, and $D(-3, 1)$ are the vertices of a figure in the coordinate plane. Plot the points and find the perimeter of the figure.

SOLUTION

Plot the points as shown in the coordinate plane at the right. Then connect the points with segments to form the figure, which is a rectangle. Notice that the sides of the figure are vertical and horizontal, so you can find the side lengths as follows.

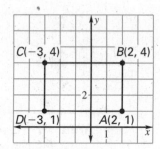

$$AB = |4 - 1| \qquad BC = |-3 - 2|$$
$$= 3 \qquad\qquad = 5$$
$$CD = |1 - 4| \qquad DA = |2 - (-3)|$$
$$= 3 \qquad\qquad = 5$$

So, the perimeter of the rectangle is

$$\text{perimeter} = 3 + 5 + 3 + 5$$
$$= 16 \text{ units.}$$

The points given are the vertices of a figure in the coordinate plane. Plot the points and find the perimeter of the figure.

1. $P(-2, 1)$, $Q(4, 1)$, $R(4, -2)$, $S(-2, -2)$

2. $W(-1, 3)$, $X(3, 3)$, $Y(3, -3)$, $Z(-1, -3)$

3. $A(-3, 0)$, $B(1, 0)$, $C(1, -4)$, $D(-3, -4)$

EXAMPLE 2

The points $A(-4, 4)$, $B(1, 4)$, $C(1, -1)$, and $D(-4, -1)$ are the vertices of a figure in the coordinate plane. Plot the points and find the area of the figure.

SOLUTION

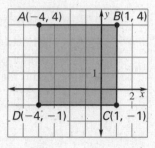

Plot the points as shown in the coordinate plane at the right. Then connect the points with segments to form the figure, which is a square. To find the area of the square, count the number of unit squares in the interior of the figure. There are 25, so the area of the figure is 25 square units.

(continued)

NAME _____ DATE _____

Perimeter and Area

The points given are the vertices of a figure in the coordinate plane. Plot the points and find the area of the figure.

4. $C(-1, 3), D(3, 3), E(3, 1), F(-1, 1)$

5. $W(-4, -1), X(-1, -1), Y(-1, -4), Z(-4, -4)$

6. $J(-1, 4), K(2, 4), L(2, -2), M(-1, -2)$

One way to find the area of a region is to divide the region into rectangles, and then add the areas of the rectangles. For instance, to find the area of the figure below, divide the figure as shown. Then find the sum of the areas of the rectangles.

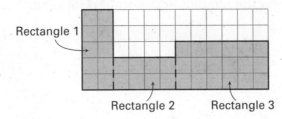

Rectangle 1

Rectangle 2   Rectangle 3

Area = Area of rectangle 1 + Area of rectangle 2 + Area of rectangle 3

= 10 + 8 + 18

= 36 square units

Another way to find the area of a region is to divide the region into areas that can be rearranged to form rectangles. For instance, the parallelogram below can be rearranged to form a rectangle.

Area of parallelogram = Area of rectangle

= 24 square units

EXAMPLE 3

Find the area of the shaded figure.

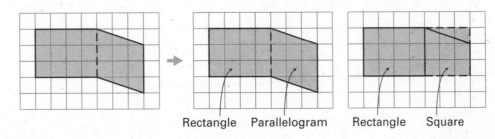

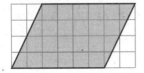Rectangle   Parallelogram   Rectangle   Square

(continued)

Perimeter and Area

SOLUTION

You can divide the figure into a rectangle and a parallelogram. The parallelogram can then be rearranged to form a square, as shown. So, the area of the figure is

Area = Area of rectangle + Area of parallelogram

    = Area of rectangle + Area of square

    = 12 + 9

    = 21 square units.

Divide and/or rearrange the shaded figure to find its area.

7.   **8.**    **9.**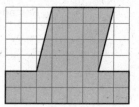

Mixed Review

10. Find the reciprocal of $\dfrac{9}{8}$.

11. Evaluate: $\dfrac{4}{15} \div \dfrac{2}{9} + \dfrac{1}{3}$.

12. Evaluate $|x - 6| + 2y$ when $x = 5$ and $y = -4$.

13. Evaluate $x^2 + y^2 - z^2$ when $x = 4$, $y = -3$, and $z = -2$.

In Exercises 14–17, use the cylinder and cone shown below.

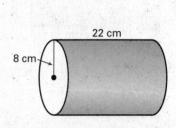
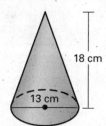

22 cm

8 cm

18 cm

13 cm

14. What is the radius of the base of the cylinder?

15. What is the height of the cylinder?

16. What is the height of the cone?

17. Which is greater, the radius of the cone's base or the diameter of the cylinder's base?

18. Simplify: $(4c - 2b) - (6b - c)$.

Geometry
Basic Skills Workbook: Diagnosis and Remediation

NAME _____ DATE _____

Quick Check

For use before Lesson 5.4: Perimeter and Area

Standardized Testing Quick Check

1. When you plot the points $(-2, 6)$, $(-2, 2)$, $(1, 2)$, and $(1, 6)$ and connect them in order from first to last, the polygon formed is a

   A. triangle.   **B.** square.

   C. rectangle.   **D.** parallelogram.

2. Triangle 1 is transformed by replacing each y with $-y$. The transformed triangle is

   A. Triangle 2.   **B.** Triangle 3.

   C. Triangle 4.   **D.** Triangle 5.

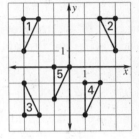

3. Which of the following transformations of a rectangle does *not* produce a rectangle that is the same size and shape?

   A. adding 2 to each x-coordinate

   B. multiplying each x-coordinate by 2

   C. subtracting 2 from each y-coordinate

   D. replacing each x-coordinate with $-x$

Homework Review Quick Check

Graph each set of points. Then connect the points in order with segments to form a polygon. Name the polygon.

4. $A(3, 1)$, $B(0, 3)$, $C(-3, 1)$, $D(-1, -2)$, $E(1, -2)$

5. $A(0, 0)$, $B(4, 2)$, $C(2, 6)$

6. Transform the polygon in Exercise 4 by subtracting 5 from each y-coordinate. Write the new coordinates of the polygon. Describe the transformation.

7. Transform the polygon in Exercise 5 by replacing each x-coordinate with $-x$. Write the new coordinates of the polygon. Describe the transformation.

Practice

For use before Lesson 5.4: Perimeter and Area

Use the figure at the right to complete the statement.

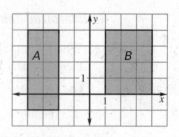

1. The perimeter of Figure *A* is _____.

2. The area of Figure *A* is _____.

3. The area of Figure *A* is _____ than the area of Figure *B*.

The points are the vertices of a figure in the coordinate plane. Plot the points and find the perimeter of the figure.

4. $A(0, 3), B(0, -1), C(-3, -1), D(-3, 3)$

5. $A(-6, 6), B(2, 6), C(2, 2), D(-6, 2)$

6. $J(8, 8), K(8, -8), L(-4, -8), M(-4, 8)$

7. $J(-6, 6), K(3, 6), L(3, -3), M(-6, -3)$

8. $W(1, -1), X(3, -1), Y(3, -2), Z(1, -2)$

9. $W(-6, -8), X(-6, 0), Y(2, 0), Z(2, -8)$

The points are the vertices of a figure in the coordinate plane. Plot the points and find the area of the figure.

10. $A(-2, 2), B(3, 2), C(3, -1), D(-2, -1)$

11. $A(-2, -3), B(-2, 2), C(1, 2), D(1, -3)$

12. $J(-1, 0), K(1, 0), L(1, -2), M(-1, -2)$

13. $J(-2, 1), K(1, 1), L(1, -3), M(-2, -3)$

14. $W(2, -2), X(-2, -2), Y(-2, 2), Z(2, 2)$

15. $W(3, 3), X(3, -1), Y(-3, -1), Z(-3, 3)$

Divide and/or rearrange the shaded figure to find its area.

16.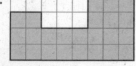

17.

18.

19. A rectangular plot of land is mapped as shown. Each square on the grid represents 100 square yards. An acre of land has an area of 4840 square yards. Is the plot of land more or less than 1 acre? Explain.

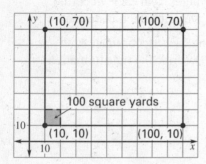

NAME _____ DATE _____

Assessment

For use with Topic 5: The Coordinate Plane

**Give the coordinates of each point on the graph and the quadrant
in which it lies.**

1. A

2. B

3. C

4. D

5. E

6. F

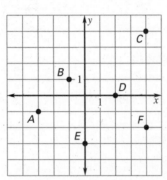

Decide whether the given ordered pair is a solution of the equation.

7. $y = -5x + 3, (-2, -7)$ 8. $3x - 2y = 10, (8, 7)$ 9. $y + 4 = -6x, (2, -1)$

Graph the linear equation.

10. $y = x + 3$

11. $x + y = 0$

12. $y = -2x - 1$

13. $4y = 2x - 8$

14. $y = 2$

15. $5x = 2y + 3$

**Graph each set of points, connect them in order with segments to
form a polygon. Then name the polygon.**

16. $(-4, -1), (-4, 5), (2, 5), (2, -1)$ 17. $(-1, -1), (1, 2), (3, -1)$

18. $(1, -3), (1, 1), (5, 1), (5, -3)$ 19. $(0, 0), (2, 0), (2, 3), (0, 3), (-1, 1)$

20. Transform the polygon in Exercise 16 by subtracting 2 from each x-coordinate.

21. Transform the polygon in Exercise 17 by replacing each y-coordinate with $-y$.

**The points are the vertices of a figure in the coordinate plane.
Plot the points and find the perimeter and area of the figure.**

22. $(6, 4), (-6, 4), (-6, -4), (6, -4)$ 23. $(-5, 4), (-2, 4), (-2, 1), (-5, 1)$

24. $(1, 4), (1, -4), (3, -4), (3, 4)$ 25. $(-3, 0), (5, 0), (5, -5), (-3, -5)$

Divide and/or rearrange the shaded figure to find its area.

26.

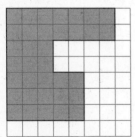

27.

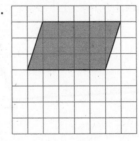

28.

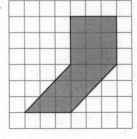

NAME _____ DATE _____

Cumulative Assessment

For use after Topics 1–5

Find the sum or difference.

1. $\left(-\frac{2}{3}\right) + \left(-\frac{2}{5}\right)$

2. $-16.75 + (-11.21)$

3. $\left(-\frac{15}{16}\right) - \left(-\frac{15}{16}\right)$

4. $1 - \left(-\frac{11}{12}\right)$

5. $\frac{13}{15} - \left(-\frac{2}{5}\right)$

6. $-2.92 + (-14.07)$

Find the product or quotient.

7. $\frac{7}{10} \div \frac{2}{5}$

8. $(-2.6)(-1.4)$

9. $(-3.27) \div (-1.09)$

10. $\left(\frac{5}{12}\right) \cdot \left(-\frac{4}{3}\right)$

11. $\left(-\frac{7}{8}\right) \div \left(-6\frac{1}{2}\right)$

12. $(5.75) \div (-1.15)$

Simplify by combining like terms.

13. $3b - 12b$

14. $8(-2b + 1) - 15$

15. $2y + 4x - 6y$

16. $26a + 3b + \frac{7}{9}b$

17. $1.2p^2 - 2.7p^2$

18. $6m^2 - 5mn + n^2 - 3mn$

Show the steps and state the mathematical properties used in simplifying the expression.

19. $-14(x - 1) + 4(2x + 1)$

20. $-8x + 3(x - 2)$

21. $4 - 12(x - 5)$

Solve the equation or inequality.

22. $16x - 12 = 36$

23. $-9 = 7 - 4d$

24. $16 - 3p \geq 5$

25. $\frac{p}{5} = -12$

26. $x - 17 \leq -4$

27. $-34 \geq -9n + 11$

Solve the proportion.

28. $\frac{c}{21} = \frac{32}{48}$

29. $\frac{8.5}{x} = \frac{17}{10}$

30. $\frac{x + 1}{x} = \frac{12}{11}$

Classify the triangle by its sides and angles. Then use a protractor to find the measure of $\angle A$ of the triangle.

31.
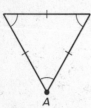

32.
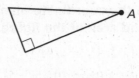

33.
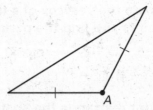

Give the name that best describes the quadrilateral.

34.

35.

36.
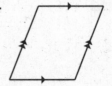

(continued)

**Graph each set of points, and connect them in order with
segments to form a polygon. Then name the polygon.**

37. $(-5, 0), (-5, 3), (1, 3), (1, 0)$ **38.** $(2, -1), (2, 5), (-4, 5), (-4, -1)$

39. Transform the polygon in Exercise 37 by subtracting 2 from each y-coordinate.

40. Find the perimeter and area of the polygon in Exercise 38.

ANSWERS

Diagnostic Test *pages v–xii*

1. 6.73 **2.** $\frac{4}{9}$ **3.** 0.33 **4.** $2\frac{5}{12}$ **5.** 0

6. $\frac{21}{16}$ **7.** 4.79 **8.** 1.13 **9.** 0 **10.** $-\frac{1}{3}$ **11.** 0

12. -15.704 **13.** -48.31 **14.** 0 **15.** $19\frac{5}{8}$

16. $\frac{2}{11}$ **17.** 8.12 **18.** $10\frac{7}{20}$ **19.** -16.22

20. $\frac{5}{6}$ **21.** 5.0 **22.** 15.08 **23.** -15.39

24. 23.25 **25.** $-\frac{1}{2}$ **26.** $\frac{5}{8}$ **27.** $-\frac{8}{9}$ **28.** 0

29. -3.81 **30.** $\frac{6}{25}$ **31.** $-\frac{67}{90}$ **32.** $42\frac{2}{3}$

33. $-1\frac{1}{13}$ **34.** $-\frac{1}{46}$ **35.** $\frac{10}{9}$ **36.** $\frac{9}{22}$ **37.** $\frac{4}{25}$

38. $\frac{5}{3}$ **39.** 1 **40.** -75.24 **41.** 40.6 **42.** $\frac{16}{21}$

43. $-2\frac{4}{9}$ **44.** -216 **45.** 100 **46.** 1 **47.** 0

48. -10 **49.** -11 **50.** $-\frac{5}{48}$ **51.** -1.672

52. $3\frac{29}{32}$ **53.** $-\frac{11}{36}$ **54.** -5 **55.** -1.54

56. $11\frac{1}{3}$ **57.** $2\frac{7}{9}$ **58.** 8.7 **59.** $18\frac{3}{4}$ **60.** -40

61. -6 **62.** 14 **63.** -2 **64.** $\frac{1}{4}$ **65.** -8

66. 25 **67.** -5 **68.** $-63\frac{7}{16}$ **69.** $8\frac{3}{4}$

70. $10x - 15$ **71.** $-7b$ **72.** $-2y - 4$

73. $-c - 9$ **74.** $30 - 11d$ **75.** $-\frac{35}{12}y$

76. $\frac{16}{9}m$ **77.** $4\frac{1}{6} + 2t$ **78.** $16x + 30$

79. $16x + y$ **80.** $3y$ **81.** $13a^2 + 2a$

82. $x^2y + 2xy - 1$ **83.** $18x^3 + 16x^2 + 3x$

84. $119x - \frac{1}{6}y$ **85.** $-\frac{1}{2}s^2t + \frac{1}{12}st^2$ **86.** $4h$

87. $-b^2 + 0.6b$ **88.** Commutative Property

89. Distributive Property

90. Addition property of zero

91. Multiplication property of zero

92. $12(x - 3) - 6(5x - 1)$
$= 12x - 36 - 30x + 6$ Distributive property
$= 12x - 30x - 36 + 6$ Commutative property
$= -18x - 30$ \qquad Simplify.

93. $\left(\frac{1}{5} + y + \frac{2}{5} - y\right)(2x + y + 8x)$
$= \left(\frac{1}{5} + \frac{2}{5} + y - y\right)(2x + 8x + y)$
\qquad Commutative property
$= \left(\frac{3}{5} + y - y\right)(10x + y)$ Simplify.
$= \left(\frac{3}{5}\right)(10x + y)$ Addition property of zero
$= 6x + \frac{3}{5}y$ Distributive property

94. $14e + f - 7e + f - 2f$
$= 14e - 7e + f + f - 2f$
\qquad Commutative property
$= 7e + 2f - 2f$ Simplify.
$= 7e$ Addition property of zero

95. $29 - 5(x - 3)$
$= 29 - 5x + 15$ Distributive property
$= 29 + 15 - 5x$ Commutative property
$= 44 - 5x$ Simplify.

96. -56 **97.** -22 **98.** -9 **99.** -3 **100.** 0

101. 0 **102.** 0.2 **103.** 42 **104.** -2

105. yes **106.** no **107.** no **108.** $x \le -12$

109. $m > -\frac{1}{5}$ **110.** $n < -7$ **111.** $x < 0$

112. $x \ge -5$ **113.** $x \le 4$ **114.** $x > -6$

115. $p > 14$ **116.** $r \ge 73$ **117.** 6 **118.** 18

119. 25 **120.** -5 **121.** 50 **122.** 20

123. $-\frac{14}{11}$ **124.** 12 **125.** 15 **126.** 90.5

127. \$12.95 **128.** $133°$, obtuse **129.** $90°$, right

130. $72°$, acute

131.   **132.**

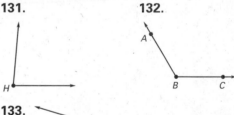

133.

134. isosceles obtuse **135.** scalene right

136. equilateral equiangular **137.** parallelogram

138. trapezoid **139.** square

140. Check drawing. diameter = 14 cm

141. Check drawing. radius = $\frac{1}{2}$ in.

142. Check drawing. diameter = 60 mm

143. $\overline{AC}, \overline{CB}, \overline{BA}, \overline{DE}, \overline{EF}, \overline{FD}, \overline{AD}, \overline{CE}, \overline{BF}$

144. A, B, C, D, E, F

145. $ABC, DEF, CEFB, ADEC, ADFB$

146. ABC, DEF **147.** cylinder **148.** sphere

149. pyramid **150.** cone **151.** prism

Diagnostic Test *continued*

152. cylinder   **153–158.**

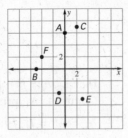

159. $(-1, 2)$, II   **160.** $(4, 0)$, none; on the *x*-axis

161. $(0, -3)$, none; on the *y*-axis   **162.** $(-2, -2)$, III

163.

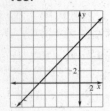

164.

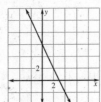

165.

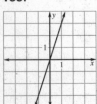

166.

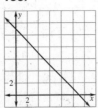

167.

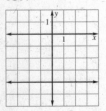

168.

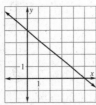

169.

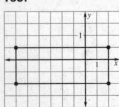

170.

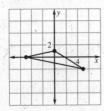

rectangle   triangle

171.

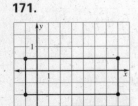

172.

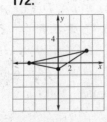

173.

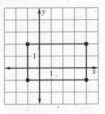

16 units,
15 square units

174.

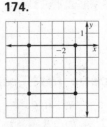

16 units,
16 square units

175.

20 units,
16 square units

176.

12 units,
8 square units

177. 34 square units   **178.** 30 square units

179. 28 square units

Topic 1
Lesson 1

Lesson Exercises *(pp. 2–3)*

1. -1 **2.** 5.71   **3.** $-7\frac{1}{5}$ **4.** 1   **5.** -7
6. 25.98   **7.** -8 **8.** -20 **9.** 25   **10.** -10

Quick Check *(p. 4)*

1. A   **2.** A   **3.** -11 **4.** -5 **5.** -35

Practice *(p.5)*

1. 9.37   **2.** $\frac{1}{7}$ **3.** 0   **4.** $3\frac{4}{9}$ **5.** $0.6\overline{1}$ **6.** $\frac{13}{5}$
7. 4.4   **8.** -4.14 **9.** 4.14   **10.** -1 **11.** 0
12. $-\frac{1}{4}$ **13.** 0   **14.** $-\frac{22}{3}$ **15.** 0   **16.** 66.101
17. -133.957 **18.** -1344.08 **19.** $-\frac{1}{2}$ **20.** $\frac{3}{56}$
21. $\frac{7}{88}$ **22.** $-16\frac{3}{4}$ **23.** $11\frac{1}{3}$ **24.** $-65\frac{1}{6}$
25. $36\frac{41}{49}$ **26.** $-1.\overline{8}$ **27.** $-\frac{17}{3}$ **28.** 4.63
29. -7 **30.** 51   **31.** 60.63   **32.** -100
33. $5\frac{1}{4}$ **34.** \$1096.75   **35.** $208\frac{1}{4}$ in.

Topic 1 *continued*

Lesson 2

Lesson Exercises *(pp. 6–7)*
1. $4 + (-5.38)$ **2.** $\left(-\frac{1}{8}\right) + \left(\frac{1}{4}\right)$
3. $\left(-10\frac{1}{11}\right) + \left(-\frac{6}{11}\right)$ **4.** $\frac{5}{8}$ **5.** 2.133 **6.** -47
7. -1 **8.** -7.81 **9.** 0 **10.** -0.48 **11.** $-\frac{9}{4}$
12. 2 **13.** $52\frac{2}{5}$ **14.** $-\frac{25}{72}$ **15.** $\$3.00$

Quick Check *(p. 8)*
1. C **2.** D **3.** -1 **4.** 23.01 **5.** $-\frac{2}{3}$

Practice *(p. 9)*
1. -11.09 **2.** $\frac{1}{4}$ **3.** 0 **4.** $-\frac{21}{2} + \left(-\frac{1}{2}\right)$
5. $8.49 + (0.07)$ **6.** $6 + (-11.976)$
7. 6.21 **8.** -6.57 **9.** 6.57 **10.** $-\frac{1}{2}$ **11.** $\frac{3}{4}$
12. $-\frac{3}{4}$ **13.** $\frac{22}{3}$ **14.** 0 **15.** 0 **16.** -2568.13
17. -51.43 **18.** 214.15 **19.** $-\frac{1}{14}$ **20.** $-\frac{61}{72}$
21. $\frac{11}{105}$ **22.** $-167\frac{1}{4}$ **23.** $11\frac{1}{9}$ **24.** $-152\frac{5}{21}$
25. $36\frac{141}{249}$ **26.** $1.\overline{8}$ **27.** $-\frac{32}{3}$ **28.** -14.3
29. $-\frac{19}{4}$ **30.** -72 **31.** $\$1500.28$ **32.** $5\frac{1}{8}$ yd

Lesson 3

Lesson Exercises *(pp. 11–13)*
1. -143.2 **2.** 6410 **3.** -0.4446 **4.** $\frac{8}{9}$
5. -11 **6.** $-\frac{8}{15}$ **7.** $-\frac{3}{20}$ **8.** -7 **9.** 0
10. 134.4 **11.** -33 **12.** -4239.48
13. -3.93 **14.** $\frac{17}{4}$ **15.** -39.072

Quick Check *(p. 14)*
1. C **2.** D **3.** $-\frac{145}{8}$ **4.** 60.454 **5.** $\frac{29}{6}$

Practice *(p. 15)*
1. $-\frac{1}{81}$ **2.** $\frac{7}{6}$ **3.** $\frac{10}{53}$ **4.** $\frac{7}{11} \cdot \left(-\frac{1}{4}\right)$
5. $\left(-\frac{1}{9}\right) \cdot (-8)$ **6.** $34\frac{1}{2} \cdot \left(\frac{2}{3}\right)$ **7.** -0.00159
8. 0.1896 **9.** -1 **10.** $\frac{2}{9}$ **11.** 2 **12.** 2
13. $-\frac{3}{8}$ **14.** $\frac{49}{25}$ **15.** 1 **16.** 352.66
17. -251 **18.** 130.1 **19.** $\frac{7}{9}$ **20.** $-5\frac{5}{54}$ **21.** 0
22. -4 **23.** -64 **24.** 4 **25.** $-26{,}717.74$
26. -8888 **27.** 100 **28.** 1 **29.** 20
30. -9 **31.** $\$.30$ **32.** 20 slices

Lesson 4

Lesson Exercises *(pp. 16–19)*
1. 1 **2.** -2.409997 **3.** 35.71 **4.** -2605
5. 15 **6.** -20.85 **7.** 177.9 **8.** -1280
9. 230.8 **10.** 70.16 **11.** 414 **12.** 3
13. $-\frac{13}{2}$ **14.** -119.75 **15.** $44\frac{4}{5}$

Quick Check *(p. 20)*
1. A **2.** C **3.** $-\frac{1}{6}$ **4.** 1.4806 **5.** -1

Practice *(p. 21)*
1. -28.1 **2.** 20.5 **3.** 3 **4.** 67 **5.** 3
6. $\frac{3}{64}$ **7.** -14.936 **8.** -13.08 **9.** 14.936
10. $\frac{33}{35}$ **11.** $\frac{6}{245}$ **12.** $1\frac{4761}{4900}$ **13.** -100
14. $-\frac{100}{81}$ **15.** $\frac{100}{81}$ **16.** -1.7999 **17.** $\frac{109}{16}$
18. $\frac{1}{4}$ **19.** 16 **20.** 15.6 **21.** 15.6
22. -1626 **23.** 69 **24.** -68.926 **25.** -10
26. $-\frac{3}{4}$ **27.** 81.67 **28.** 4793.7 **29.** $-\frac{200}{3}$
30. $\frac{407}{2}$ **31.** $\$10.91$

Assessment *(p. 22)*
1. 8.5 **2.** $\frac{1}{3}$ **3.** $4\frac{3}{8}$ **4.** -1 **5.** 75 **6.** -25
7. 1 **8.** $-\frac{16}{15}$ **9.** -8.31 **10.** -4.08
11. $\frac{2}{15}$ **12.** 0 **13.** $\frac{3}{10}$ **14.** $-\frac{44}{21}$
15. -22.66 **16.** $-\frac{1}{23}$ **17.** $\frac{7}{6}$ **18.** 4 **19.** $\frac{1}{6}$
20. -19.32 **21.** $-\frac{5}{66}$ **22.** 131 **23.** $2\frac{4}{9}$
24. 3.75 **25.** $-\frac{20}{7}$ **26.** -70.956 **27.** $\frac{3}{10}$
28. 9 **29.** $\frac{3}{20}$ **30.** 4 **31.** 4 **32.** 0.125 **33.** 27
34. $\frac{3}{2}$ **35.** $\frac{17}{3}$ **36.** $\frac{250}{29}$

Topic 2
Lesson 1

Lesson Exercises *(pp. 24–26)*
1. -7 **2.** 30 **3.** $\frac{13}{10}$ **4.** $-\frac{9}{8}$ **5.** -1 **6.** $9\frac{1}{8}$
7. 7 **8.** $-\frac{63}{440}$ **9.** 3.5 **10.** 3 in.^{2}
11. about 14.44 m **12.** 100 ft^2 **13.** -49.9
14. 1.75 **15.** 1

Quick Check *(p. 27)*
1. C **2.** D **3.** 1.24 **4.** $\frac{3}{4}$ **5.** 3.18

Topic 2 continued

Practice (p. 28)

1. -4 **2.** 2 **3.** 2 **4.** -1 **5.** $\frac{1}{3}$ **6.** -6
7. $\frac{328}{9}$ **8.** 12 **9.** 3 **10.** $\frac{11}{3}$ **11.** 25 **12.** 17
13. 64 **14.** $\frac{20}{3}$ **15.** $-\frac{14}{3}$ **16.** 99.07
17. -7.07 **18.** -7.07 **19.** 0.07 **20.** 100.93
21. 101.07 **22.** -0.9993 **23.** -0.0093
24. 1 **25.** -78 **26.** 10,001.0049 **27.** -2.72
28. -6.51 **29.** -7 **30.** 8.07 **31.** 22.4 m
32. about 7.065 in.^{2} **33.** $13\frac{3}{5}$ cm **34.** 37.5 km
35. 96 ft^2

Lesson 2

Lesson Exercises (p. 30)

1. $11b + 0.6$ **2.** 0 **3.** $5.3c + 2.9$ **4.** $-0.8x$
5. $4x + 12$ **6.** $-10x - 8$ **7.** 34
8. 10.65 in. **9.** 7 **10.** -11

Quick Check (p. 31)

1. A **2.** C **3.** -24 **4.** 361 **5.** $\frac{3}{64}$

Practice (p. 32)

1. no **2.** yes **3.** yes **4.** yes **5.** no **6.** no
7. $15x$ **8.** $-2y$ **9.** $11a + 17$ **10.** $7c - 10$
11. $7d + 15$ **12.** $66h$ **13.** $11b$
14. $-5y - 2$ **15.** $-x + 14$ **16.** $6.1x - 0.94$
17. $-5a + 8$ **18.** $21x + 2504$ **19.** k
20. $\frac{7}{2}n + 1$ **21.** $\frac{2}{7}y + 60$ **22.** $10x + 14$
23. $-86.24z + 6.44$ **24.** $28n - 88$ **25.** $\frac{3}{2}y$
26. $-\frac{21}{4}v$ **27.** $5c + 5\frac{7}{8}$ **28.** $-13.43x + 7$
29. $7d - 854.3$ **30.** $-16.08t + 21$ **31.** $6x$
32. $12a - 2$

Lesson 3

Lesson Exercises (pp. 33–34)

1. $-8a + 4b$ **2.** $75x^2$ **3.** $16cd - 2$
4. $40s^2 - 23s + 3$ **5.** $a^2 + 5ab + 2b^2$
6. $x - 4y$ **7.** $10.38z$ **8.** $20x - 28$ **9.** -9
10. $-14,757.12$

Quick Check (p. 35)

1. B **2.** C **3.** $92a + 60$ **4.** 0 **5.** $5x$

Practice (p. 36)

1. $5x, x$ **2.** $-7xy, xy$
3. $11a^3, 20a^3; -a^2, -9a^2$ **4.** $4b, 3b$ **5.** none
6. $3xy, xy$ **7.** $73x + y$ **8.** $-4y$
9. $32a^2 + 6a$ **10.** $7c - 20d$
11. $105 + cd + 6d$ **12.** $5m + 13n$
13. $33b^2 - 2b$ **14.** $3c^2$ **15.** $-3x - 4y$
16. $x^2y + 9xy - 1$ **17.** $-8m + 3.33n$
18. $59x^3 + 45x^2 + 21x$ **19.** $\frac{4}{3}k + \frac{1}{6}l$
20. $172x - \frac{4}{9}y$ **21.** $9m^2 - 4mn - 4n^2$
22. $-6.9r^2$ **23.** $-5373.73z + 0.165$
24. $52.17x^2 + 0.24x$ **25.** $\frac{17}{16}xy + \frac{35}{4}$
26. $-\frac{5}{2}u^3v - \frac{5}{4}uv^3$ **27.** $64 + c + d$ **28.** 0
29. $5A + 2B$ **30.** $-2t^2 + 2.04t$ **31.** $24x$

Lesson 4

Lesson Exercises (pp. 37–38)

1. **a.** Distributive property
   b. Commutative prop. of addition
   c. Addition prop. of zero
   d. Identity prop. of addition
   e. Identity prop. of mult.

2. **a.** Associative prop. of addition
   b. Distributive property
   c. Distributive property
   d. Commutative prop. of addition
   e. Distributive property

3. $11k^2 - 15k + 8(k^2 + 2k)$

$= 11k^2 - 15k + 8k^2 + 16k$

   Distributive prop.

$= 11k^2 + 8k^2 - 15k + 16k$

   Commutative prop. of add.

$= (11 + 8)k^2 + (-15 + 16)k$

   Distributive prop.

$= 19k^2 + 1k$

   Simplify.

$= 19k^2 + k$

   Identity prop. of mult.

Topic 2 *continued*

4. $9[(a^2 + ab) + 2ab] - 3(9ab + b^2)$

$\quad = 9[a^2 + (ab + 2ab)] - 3(9ab + b^2)$

$\qquad\qquad$ Assoc. prop. of add.

$\quad = 9(a^2 + 3ab) - 3(9ab + b^2)$

$\qquad\qquad$ Simplify.

$\quad = 9a^2 + 27ab - 27ab - 3b^2$

$\qquad\qquad$ Distributive prop.

$\quad = 9a^2 - 3b^2$

$\qquad\qquad$ Add. prop. of zero

5. $[-3 - (x - 3)](2x + y - y - 2x)$

$\quad = (-3 - x + 3)(2x + y - y - 2x)$

$\qquad\qquad$ Distributive prop.

$\quad = (-3 + 3 - x)(2x - 2x + y - y)$

$\qquad\qquad$ Commutative prop. of add.

$\quad = (-x)(0)\qquad$ Addition prop. of zero

$\quad = 0\qquad\qquad$ Mult. prop. of zero

6. -1 **7.** $2x^2$ **8.** 10.18 **9.** $\frac{1}{8}$

Quick Check *(p. 39)*

1. C **2.** A **3.** $27 - \frac{1}{2}c^2 - \frac{1}{2}d^2$

4. $m^5n + 14mn^5$ **5.** $2x^2 - 100.02y^2$

Practice *(p. 40)*

1. Comm. prop. of add. **2.** Dist. prop.

3. Add. prop. of zero **4.** Assoc. prop. of addition **5.** Identity prop. of mult.

6. Mult. prop. of zero **7. a.** Comm. prop. of add. **b.** Add. prop. of zero **c.** Identity prop. of add. **d.** Identity prop. of mult. **8. a.** Comm. prop. of add. **b.** Distributive prop. **c.** Assoc. prop. of add. **d.** Add. prop. of zero

9. a. Comm. prop. of add. **b.** Assoc. prop. of add. **c.** Add. prop. of zero **d.** Identity prop. of add.

10. $14(x - 1) - 7(2x - 2)$

$\quad = 14x - 14 - 14x + 14$

$\qquad\qquad$ Dist. prop.

$\quad = 14x - 14x - 14 + 14$

$\qquad\qquad$ Commutative prop. of add.

$\quad = 0\qquad\qquad$ Add. prop. of zero

11. $\left(\frac{1}{2} + x + \frac{1}{2} - x\right)(3x + y + 2x)$

$\quad = \left(\frac{1}{2} + \frac{1}{2} + x - x\right)(3x + 2x + y)$

$\qquad\qquad$ Comm. prop. of add.

$\quad = (1 + x - x)(5x + y)$

$\qquad\qquad$ Simplify.

$\quad = (1)(5x + y)$

$\qquad\qquad$ Add. prop. of zero

$\quad = 5x + y\qquad$ Identity prop. of mult.

12. $[7uv + (v - 7uv)][(u + v) - (u + v)]$

$\quad = [(7uv - 7uv) + v][(u + v) - (u + v)]$

$\qquad\qquad$ Comm. and Assoc. prop. of add.

$\quad = (v)(0)\quad$ Add. prop. of zero

$\quad = 0\quad$ Mult. prop. of zero

13. $16 - 5(x + 4) = 16 - 5x - 20$

$\qquad\qquad$ Distributive prop.

$\quad = 16 - 20 - 5x$

$\qquad\qquad$ Comm. prop. of addition

$\quad = -4 - 5x$

$\qquad\qquad$ Simplify.

14. $11c + d + c + d - 2d$

$\quad = 11c + c + d + d - 2d$

$\qquad\qquad$ Comm. prop. of addition

$\quad = 12c + 2d - 2d$

$\qquad\qquad$ Simplify.

$\quad = 12c\qquad$ Addition prop. of zero

15. $(g + 6h) + (43h + 23g) + 89g$

$\quad = (g + 6h) + 43h + (23g + 89g)$

$\qquad\qquad$ Assoc. prop. of addition

$\quad = (g + 6h) + 43h + 112g$

$\qquad\qquad$ Simplify.

$\quad = g + (6h + 43h) + 112g$

$\qquad\qquad$ Assoc. prop. of addition

$\quad = g + 49h + 112g$

$\qquad\qquad$ Simplify.

$\quad = g + 112g + 49h$

$\qquad\qquad$ Comm. prop. of addition

$\quad = 113g + 49h$

$\qquad\qquad$ Simplify.

Answers

Topic 2 *continued*

Assessment *(p. 41)*

1. -9 2. $6\frac{1}{2}$ 3. -60 4. 46 5. 32 6. $1\frac{1}{2}$

7. $-17x$ 8. $-9b + 10$ 9. $1.9a - 7$

10. 0 11. $2\frac{1}{3} + \frac{4}{3}x$ 12. $-21y + 4$

13. $-2y + 4x$ 14. $4xy - 5y$

15. $10x^2y^2 + 4x^2y$ 16. $23x - \frac{11}{8}y$ 17. $-2.3a^2$

18. $-8m^2 - 2mn - n^2$

19. Commutative prop. of addition

20. Distributive property

21. Assoc. prop. of addition

22. Identity prop. of multiplication

23. a. Distributive property

    b. Commutative prop. of addition

    c. Add. prop. of zero

24. $[2x - (y + 2x)][(-3 + x) - x + 4]$

    $= [2x - y - 2x][(-3 + x) - x + 4]$

    　　　　　　　　Distributive prop.

    $= [(2x - 2x) - y][-3 + (x - x) + 4]$

    　　　　　　　　Comm. and Assoc.
    　　　　　　　　prop. of addition

    $= (-y)(-3 + 4)$　　Add. prop. of zero

    $= (-y)(1)$　　　　Simplify.

    $= -y$　　　　　　Identity prop. of mult.

25. $-16(x - 2) + 8(2x - 4)$

    $= -16x + 32 + 16x - 32$

    　　　　　　　　Distributive prop.

    $= -16x + 16x + 32 - 32$

    　　　　　　　　Comm. prop. of addition

    $= 0$　　　　　　Addition prop. of zero

26. 66 m 27. about 67 mi/h 28. $7.5a + 11.5$

Topic 3
Lesson 1

Lesson Exercises *(pp. 43–45)*

1. 1 2. $\frac{1}{2}$ 3. -5 4. $\frac{8}{7}$ 5. -150 6. -1

7. 1.56 8. 6 9. -45 10. $\frac{1}{2}$ 11. -3

12. $-\frac{17}{7}$ 13. $-2 - x + y$ 14. 2

15. $28x - 12$

Quick Check *(p. 46)*

1. C 2. B 3. Commutative property of addition 4. Addition property of zero

5. Associative property of addition

Practice *(p. 47)*

1. Subtraction property of equality 2. Division property of equality 3. Multiplication property of equality 4. Addition property of equality, Division property of equality 5. -32 6. -12

7. 4 8. $-\frac{1}{2}$ 9. 1 10. -1 11. 0 12. 2

13. 0 14. $\frac{25}{36}$ 15. $\frac{5}{4}$ 16. 4.07 17. -1

18. 44 19. $\frac{7}{10}$ 20. $l = 8$ ft, $w = 4$ ft

Lesson 2

Lesson Exercises *(pp. 49–51)*

1. $x > -\frac{3}{10}$ 2. $d \le 112$ 3. $y \le \frac{23}{15}$

4. $x > -33$ 5. $x < 1$ 6. $y \ge -1$

7. $a < 0$ 8. $b \ge 0$ 9. $\frac{59}{10}$ 10. $3x^6 + 5x^5$

11. $\frac{3}{4}$ 12. 2 13. -1 14. $\frac{8}{5}$ 15. $27d + 55$

Quick Check *(p. 52)*

1. C 2. B 3. $-\frac{1}{3}$ 4. 2 5. -7

Practice *(p. 53)*

1. yes 2. yes 3. no 4. no 5. $x \le -3$

6. $d > -4$ 7. $d < 2$ 8. $d > -\frac{1}{3}$

9. $x \le -2$ 10. $y < -\frac{17}{13}$ 11. $k \le 7$

12. $b > 2$ 13. $x < 0$ 14. $h \ge 49$ 15. $x > \frac{1}{3}$

16. $n < 0.03$ 17. $t < -20$ 18. $x \le -\frac{3}{8}$

19. $x < -7$ 20. $k > \frac{15}{77}$ 21. $y > 50$

22. $x > 90$ 23. $a > -53$ 24. $m \le 18$

25. $x \ge 2$

Lesson 3

Lesson Exercises *(pp. 55–56)*

1. $\frac{1}{4}$ 2. 9 3. -2 4. $-\frac{24}{5}$ 5. 9 6. $-\frac{12}{5}$

7. $-\frac{1}{6}$ 8. 3 9. -75 10. $v < -1.3$

11. -1

Quick Check *(p. 57)*

1. C 2. D 3. $x < -19.5$ 4. $x \le \frac{20}{3}$

5. $b < 3$

Topic 3 *continued*

Practice *(p. 58)*

1. extremes: 2, 3; means: 6, 1   **2.** yes   **3.** no
4. yes   **5.** 4   **6.** 16   **7.** $\frac{16}{11}$ **8.** $\frac{11}{8}$ **9.** 25
10. 63   **11.** $\frac{6}{5}$ **12.** 80   **13.** 4   **14.** -33
15. -42 **16.** 5   **17.** -3 **18.** $\frac{87}{80}$ **19.** $-\frac{15}{58}$
20. 2   **21.** -12 **22.** 0   **23.** 12 m   **24.** 8 cups
25. about 0.692 mg

Lesson 4

Lesson Exercises *(pp. 60–61)*

1. 75 brownies   **2.** $14.13   **3.** 21 points
4. 10 pizzas   **5.** about 50 cups   **6.** 1
7. 37   **8.** $c < -\frac{9}{10}$ **9.** 757.83
10. $37a + 61b$ **11.** $\frac{8}{5}$ **12.** 13.6   **13.** 10

Quick Check *(p. 62)*

1. D   **2.** C   **3.** 2   **4.** 68   **5.** $\frac{700}{3}$ **6.** $-\frac{1}{5}$
7. -10 **8.** -10

Practice *(p. 63)*

1. $15.00   **2.** 31 weeks   **3.** 3 gallons
4. 60 kites   **5.** about 59 flowers   **6.** 8 2-point
field goals   **7.** about 16 tiles   **8.** 71   **9.** 7 fish
10. more than $2.50 per ticket

Assessment *(p. 64)*

1. 2   **2.** 4   **3.** -88 **4.** $-\frac{1}{4}$ **5.** 4   **6.** 2
7. -1.5 **8.** 1   **9.** 21   **10.** no   **11.** yes
12. no   **13.** $n < 4$ **14.** $k \ge 0.6$ **15.** $c > -4$
16. $y \le -3$ **17.** $d \ge 3$ **18.** $h \le -25$
19. $r > 16$ **20.** $b > 20$ **21.** $c > -2$
22. 15   **23.** 20   **24.** 1   **25.** 48   **26.** 10
27. 36   **28.** 7.2   **29.** 24   **30.** 3   **31.** 3 points
32. 18 candles   **33.** $25.50

Topic 4
Lesson 1

Lesson Exercises *(p. 67)*

1. 75°, acute   **2.** 155°, obtuse   **3.** 15°, acute
4. 180°, straight

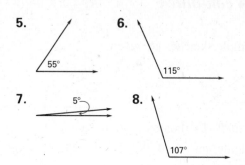

9. -1 **10.** $15v$ **11.** 40   **12.** $\frac{500}{7}$ **13.** $\frac{3}{5}$
14. $x \le 2$ **15.** $-\frac{15}{4}$

Quick Check *(p. 68)*

1. A   **2.** C   **3.** 900 students   **4.** at least 154
more votes   **5.** 2 hours per week

Practice *(p. 69)*

1. 125°, obtuse   **2.** 30°, acute   **3.** 180°, straight
4. 160°, obtuse   **5.** 90°, right   **6.** 77°, acute

7.

8.

9.

10.

11.

12.

13.

14.

Lesson 2

Lesson Exercises *(pp. 71–72)*

1.

2.

Answers

Topic 4 *continued*

3, 4. Sample sketches are given.

3.    **4.**

5. trapezoid   **6.** rhombus   **7.** rectangle

8. parallelogram   **9.** 4096   **10.** $\frac{20}{3}$ **11.** 50,490

12. $x + 7.1xy + 1.286$ **13.** $\frac{1}{2}$ **14.** $-\frac{1}{3}$ **15.** 17

Quick Check *(p. 73)*

1. B   **2.** D   **3.** 135°, obtuse   **4.** 90°, right

5. 30°, acute

Practice *(p. 74)*

1. right scalene   **2.** equiangular (acute) equilateral

3. obtuse isosceles   **4.** false   **5.** false   **6.** true

7. false   **8.** true

9. 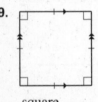   **10.** not possible

square

11. *Sample answer:*   **12.** *Sample answer:*

   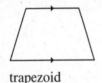

rectangle   trapezoid

13. not possible   **14.** not possible

15. *Sample answer:*

Lesson 3

Lesson Exercises *(p. 76)*

1–4. Check drawings.   **5.** 180°

6. $\frac{2997}{1000}$ **7.** $x \le -0.0495$

8. $4x + 5$ **9.** $-\frac{6}{11}$ **10.** 43°

Quick Check *(p. 77)*

1. E   **2.** B   **3.** trapezoid

4. isosceles triangle   **5.** rectangle

Practice *(p. 78)*

1. 20 mm   **2.** 34 in.   **3.** 4 ft   **4.** $\frac{12}{7}$ m

5. 10.6 cm   **6.** $18\frac{1}{8}$ in.   **7.** 9 ft   **8.** 3 in.

9. 5.5 miles   **10.** $\frac{3}{8}$ in.   **11.** 0.05 m

12. $22\frac{3}{16}$ in.   **13–22.** Check drawings.

Lesson 4

Lesson Exercises *(pp. 80–81)*

1. *ABCDE, ABIJ, BIHC, CHGD, DEFG, AEFJ,*
and *JFGHI*   **2.** *A, B, C, D, E, F, G, H, I,* and *J*

3. $\overline{AB}, \overline{BC}, \overline{CD}, \overline{DE}, \overline{EA}, \overline{AJ}, \overline{JI}, \overline{IB}, \overline{IH}, \overline{HC}, \overline{HG},$
$\overline{GD}, \overline{GF}, \overline{FJ},$ and $\overline{EF}$ **4.** polyhedron; 9 faces, 9
vertices, 16 edges   **5.** not a polyhedron

6. polyhedron; 8 faces, 12 vertices, 18 edges

7. pyramid; yes   **8.** cone; no   **9.** prism; yes

10. two   **11.** 90, 180   **12.** never

13. sometimes

Quick Check *(p. 82)*

1. C   **2.** A   **3–5.** Check drawings.

Practice *(p. 83)*

1. $\overline{DE}, \overline{EF}, \overline{FB}, \overline{BA}, \overline{AD}, \overline{AC}, \overline{CB}, \overline{DF}, \overline{CE}$

2. *A, B, C, D, E, F*   **3.** *ACED, CBFE, ABFD,*
ABC, DEF   **4.** *DEF, ACB*   **5.** 7 faces, 10 vertices, 15 edges   **6.** 8 faces, 12 vertices,
18 edges   **7.** 8 faces, 6 vertices, 12 edges

8. 8 faces, 12 vertices, 18 edges   **9.** polyhedron

10. prism   **11.** pyramid   **12.** cylinder

Assessment *(pp. 84–85)*

1. 90°, right   **2.** 40°, acute   **3.** 115°, obtuse

4. 108°, obtuse   **5.** 180°, straight

6. 20°, acute

7.   **8.**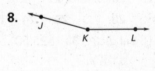

9.   **10.** equilateral, equiangular
(acute)

11. isosceles, obtuse

12. scalene, right

13. parallelogram   **14.** rhombus   **15.** trapezoid

Topic 4 *continued*

16. Check drawings. diameter = 8 cm
17. Check drawings. radius = 1 in.
18. Check drawings. diameter = 7 in.
19. pyramid **20.** cylinder **21.** prism

Topic 5
Lesson 1

Lesson Exercises *(pp. 87–88)*

1.
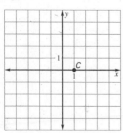

2.
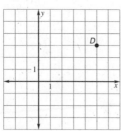

3.
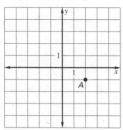

4.
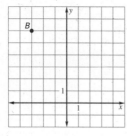

5. (3, 2); Quadrant I **6.** (0, −4); no quadrant, on axis **7.** (−2, 2); Quadrant II **8.** (4, −3); Quadrant IV **9.** −5488 **10.** 3800 **11.** 0 **12.** 0

Quick Check *(p. 89)*

1. E **2.** D **3.** sphere **4.** cylinder **5.** prism

Practice *(p. 90)*

1–10.
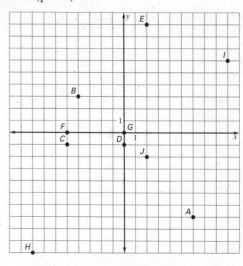

11. (8, 7); Quadrant I **12.** (7, 0); no quadrant, on axis **13.** (−5, 3); Quadrant II
14. (−2, −3); Quadrant III **15.** (8, −4); Quadrant IV **16.** (−7, 7); Quadrant II
17. (2, 6); Quadrant I **18.** (0, −8); no quadrant, on axis **19.** (−7, −5); Quadrant III

Lesson 2

Lesson Exercises *(pp. 92–93)*

1. no **2.** yes **3.** yes **4.** no

5.
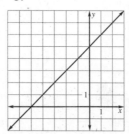

6.
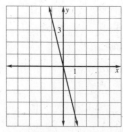

7.
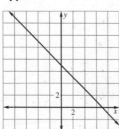

8.
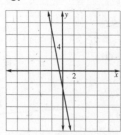

9. about 4 hours **10.** $-\frac{14}{27}$ **11.** 1

Topic 5 *continued*

Quick Check *(p. 94)*

1. D **2.** C

3.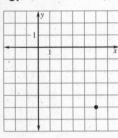

4.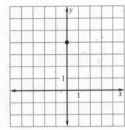

5.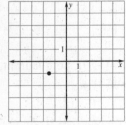

Practice *(p. 95)*

1. no **2.** no **3.** no **4.** yes

5. *Sample answer:* $(0, 12), (-1, 14), (2, 8)$

6. *Sample answer:* $(0, -7), (1, -11), (-2, 1)$

7.

x	$y = 6x + 2$	y
-2	$y = 6(-2) + 2$	-10
-1	$y = 6(-1) + 2$	-4
0	$y = 6(0) + 2$	2
1	$y = 6(1) + 2$	8
2	$y = 6(2) + 2$	14

8.

x	$y = \frac{2}{7}x - 4$	y
-14	$y = \frac{2}{7}(-14) - 4$	-8
-7	$y = \frac{2}{7}(-7) - 4$	-6
0	$y = \frac{2}{7}(0) - 4$	-4
7	$y = \frac{2}{7}(7) - 4$	-2
14	$y = \frac{2}{7}(14) - 4$	0

9.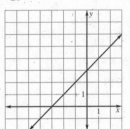

10.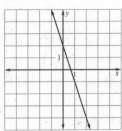

11.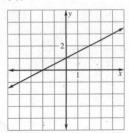

12.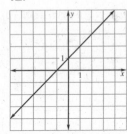

13.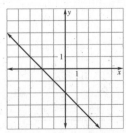

14.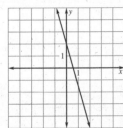

15.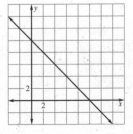

16.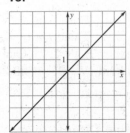

17.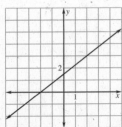

18.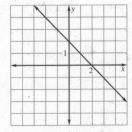

Geometry
Basic Skills Workbook: Diagnosis and Remediation

Topic 5 *continued*

19.

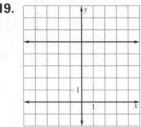

Lesson 3

Lesson Exercises *(pp. 96–98)*

1.

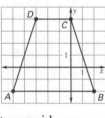

(isosceles right) triangle

2.

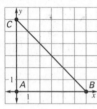

square

3.

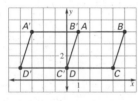

trapezoid

4.

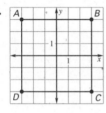

pentagon

5.

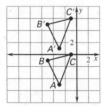

The new parallelogram is shifted 4 spaces to the left of the original parallelogram. It is the same size and shape.

6.

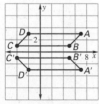

The new triangle is shifted 6 spaces up from the original triangle. It is the same size and shape.

7.

The new parallelogram is the original parallelogram flipped over the *x*-axis. It is the same size and shape.

8.

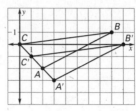

The new triangle is shifted one space to the right and one space down from the original triangle. It is the same size and shape.

9. rhombus **10.** acute angle **11.** equilateral triangle **12.** square **13.** equiangular triangle **14.** square **15.** straight angle

Quick Check *(p. 99)*

1. C **2.** C **3.**

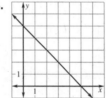

4.

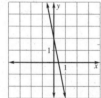

5.

Practice *(p. 100)*

1.

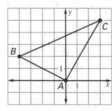

triangle

2.

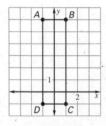

rectangle

3.

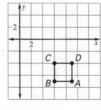

square

4.

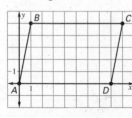

parallelogram

Topic 5 *continued*

5.

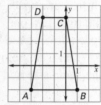

trapezoid

6.

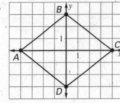

rhombus

7.

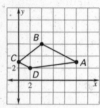

trapezoid

8.

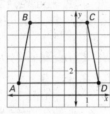

trapezoid

9.

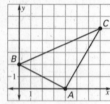

The new triangle is shifted 4 spaces to the right of the original triangle. It is the same size and shape.

10.

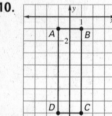

The new rectangle is shifted 7 spaces down from the original rectangle. It is the same size and shape.

11.

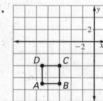

The new square is the original square flipped over the *y*-axis. It is the same size and shape.

12.

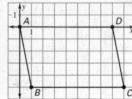

The new parallelogram is the original parallelogram flipped over the *x*-axis. It is the same size and shape.

13.

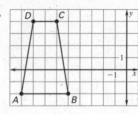

The new trapezoid is shifted 6 spaces to the left of the original trapezoid. It is the same size and shape.

14.

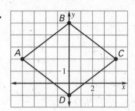

The new rhombus is shifted two spaces up from the original rhombus. It is the same size and shape.

15.

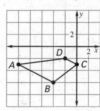

The new trapezoid is the original trapezoid flipped over the *y*-axis then flipped over the *x*-axis. It is the same size and shape.

16.

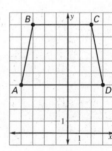

The new trapezoid is shifted one space to the right and three spaces up from the original trapezoid. It is the same size and shape.

For Exs. 17–20. Answers will vary; sample answers are given.

17.

18.

19.

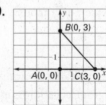

20.

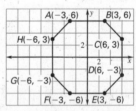

Geometry
Basic Skills Workbook: Diagnosis and Remediation

Topic 5 continued

22.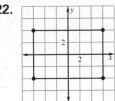

40 units,
96 square units

23.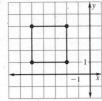

12 units,
9 square units

24.

20 units,
16 square units

25.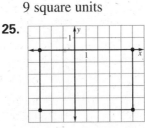

26 units,
40 square units

26. 35 square units   **27.** 15 square units

28. 18 square units

37.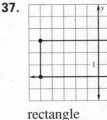

rectangle

38.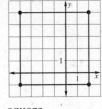

square

39.

40. 24 units,
36 square units

Cumulative Assessment *(pp. 108–109)*

1. $-1\frac{1}{15}$ **2.** -27.96 **3.** 0   **4.** $1\frac{11}{12}$ **5.** $1\frac{4}{15}$

6. -16.99 **7.** $1\frac{3}{4}$ **8.** 3.64   **9.** 3   **10.** $-\frac{5}{9}$

11. $\frac{7}{52}$ **12.** -5 **13.** $-9b$ **14.** $-16b - 7$

15. $4x - 4y$ **16.** $26a + \frac{34}{9}b$ **17.** $-1.5p^2$

18. $6m^2 - 8mn + n^2$

19. $-14(x - 1) + 4(2x + 1)$

$= -14x + 14 + 8x + 4$ Distributive prop.

$= -14x + 8x + 14 + 4$ Commutative prop.

$= -6x + 18$ Simplify.

20. $-8x + 3(x - 2)$

$= -8x + 3x - 6$ Distributive property

$= -5x - 6$ Simplify.

21. $4 - 12(x - 5)$

$= 4 - 12x + 60$ Distributive property

$= 4 + 60 - 12x$ Commutative property

$= 64 - 12x$ Simplify.

22. 3   **23.** 4   **24.** $p \le \frac{11}{3}$ **25.** -60

26. $x \le 13$ **27.** $n \ge 5$ **28.** 14   **29.** 5

30. 11   **31.** equilateral, equiangular and acute, 60°

32. scalene, right, 24°   **33.** isosceles, obtuse, 118°

34. trapezoid   **35.** square   **36.** parallelogram